Genetic Engineering

Genetic Engineering

Kerby
Anderson

ZONDERVAN
PUBLISHING HOUSE OF THE ZONDERVAN CORPORATION
GRAND RAPIDS, MICHIGAN 49506

GENETIC ENGINEERING
Copyright © 1982 by The Zondervan Corporation
Grand Rapids, Michigan

First printing, June 1982

Library of Congress Cataloging in Publication Data

Anderson, J. Kerby.
   Genetic engineering.

   Bibliography: p.
   Includes index.
   1. Genetic engineering—Religious aspects. I. Title.
QH442.A53   261.5'6   82-1880
ISBN 0-310-45051-9   AACR2

Edited by John Danilson and Gerard Terpstra
Designed by Louise Bauer

Printed in the United States of America

To my parents,
John and Lorraine Anderson,
whose nurture and encouragement
have made me much of what I am
as a writer and speaker

Contents

Preface

The age of genetics has arrived. A discussion of genetic engineering is often as current as the daily news. Christians unfortunately have often been the last to speak clearly on the issues involved. There are few books by Evangelicals concerning genetic engineering, and fewer still that seek to bring a balanced appraisal of these advances in reproductive biology and genetic research. It is therefore appropriate that this book concentrate on the various scientific, social, legal, ethical, and theological issues concerning genetic engineering.

Despite the adverse publicity and doomsday scenarios created by newspaper columnists, authors, and movie producers, it is unlikely that genetic engineering will be forbidden. Like any technology, it can bear both good and bad fruits. Christians must seek to bring a careful critique to this growing area of science. The discussion will not be advanced by the frequent diatribes made against genetic engineering. We must think, interact, discuss, and write cogently on this topic. If we do not, we have only ourselves to blame for any wrong direction genetic engineering may take.

Acknowledgments

I especially want to thank my friend and colleague Ray Bohlin for the helpful insight he gave me as I was writing this book. As a fellow speaker on this subject, he has contributed invaluable help.

I would also like to thank my patient wife for her willingness to give up frequent evening and weekend times together so that I could finish typing and editing the manuscript.

Genetic Engineering

I
Genetics
and
Society

Welcome to the age of genetics. Although many of us may not be aware of it, we are in a genetic revolution. Knowledge in genetics is doubling every few years, and we are presently undergoing a revolution in the field of genetics not unlike the nuclear revolution and the computer revolution of previous decades. Genetic engineering is no longer science fiction but is science fact. The future holds great promise as well as cause for great concern.

Genetics is at the very basis of our lives. Every living thing is tied to its genetic structure. Nearly all that we are physically and part of what we are mentally are determined by our genetic structure. It is not surprising, therefore, that mankind has always had an interest in genetic engineering. Genetic research offers a means by which we can eventually cure the many genetic diseases that afflict our population. Genetic research in agriculture, animal husbandry, and forestry has brought great blessings to us in terms of food and natural-resource production.

The future of genetic technology, like any other technology, offers great promise but also great peril. Nuclear technology has given us nuclear medicine, nuclear energy, and nuclear weapons. Genetic technology will most likely give a similar array of good, questionable, and bad technological applications. We hold in our hands the potential to solve many problems but also the potential to create many more.

This technology is as powerful as any that mankind has ever possessed. For the first time in human history it is possible to completely redesign existing organisms, including man, and direct the genetic and reproductive constitution of every living thing. No longer are we merely limited to breeding and cross-pollination. Powerful genetic tools allow us to change genetic structure at the microscopic level and by-pass the normal processes of reproduction.

In this book we will be looking at a variety of different aspects of genetic manipulation and artificial reproduction. Therefore the term *genetic engineering* will be used in its broadest sense. It refers to nearly any attempt to modify the structure, transmission, expression, or effects of any living organism's genetic structure. It includes all aspects of genetic research, genetic manipulation, and genetic breeding and reproductive biology. The following medical definition is appropriate for the genetic engineering of humans:

> The popular term, genetic engineering, might be considered as covering anything having to do with the manipulation of the gametes (reproductive cells: sperm and eggs) or the fetus, for whatever purpose, from conception other than by sexual union, to treatment of disease *in utero*, to the ultimate manufacture of a human being to exact specifications.[1]

Before dealing with specific issues, we should briefly survey in this chapter some basic genetic information that will provide an important foundation for the rest of our discussions. A glossary is provided at the back of this book for further information beyond this basic survey of the genetics of life.

THE GENETICS OF LIFE

The basic unit of inheritance is the *gene*. The gene is composed of DNA (deoxyribonucleic acid), which is a single strand of small subunits (nucleotides) arranged in a particular sequence. Each gene codes for the production of a particular protein. A protein is a molecule composed of subunits called amino acids. The genetic sequence of nucleotides is "read" and then "translated" to produce a sequence of amino acids that forms a protein.

If the DNA comprising a gene should undergo a structural alteration, this is called a *mutation*. The gene will then code for a slightly different protein or else produce a nonsense substance.

For example, sickle-cell anemia is caused by the mutation of one gene that codes for one of the amino acids in the protein hemoglobin. There are actually four proteins or polypeptides (each composed of about one hundred and forty amino acids) that make up hemoglobin. In this mutation there is the substitution of just one different amino acid thereby causing sickle-cell disease. Most *genetic diseases* are caused by minor genetic variations. Although most medical break-throughs concerning genetic disease have merely enabled a person to better live with his disease, the major goal of genetic research is to *cure* these diseases.

In a living cell, genes are located on *chromosomes*, which are double strands of DNA twisted and coiled into a double helix and surrounded by proteins. The word literally means "colored body," and chromosomes were so named because they appeared as dark bodies when first observed under a light microscope. Every organism has a specific number of chromosomes made up of matched pairs called *homologues.* Humans usually have forty-six chromosomes in each cell. Twenty-two pairs are called somatic chromosomes, and the remaining pair constitutes the sex chromosomes. A human male has an X and a Y sex chromosome, while a female has two X chromosomes.

Reproduction takes place when the chromosomes divide during meiosis to form the germ cells. The sperm or egg produced is *haploid.* In other words it carries half of the full complement of chromosomes (twenty-three in humans). When the egg and sperm merge, the haploid chromosomes join together to form the normal number of chromosomes (forty-six in humans).

Genetic diseases arise from a number of different causes.

Many are caused by a *single-gene defect*. Some of these single-gene diseases are dominant and therefore cannot be masked by a second normal gene on the other homologous chromosome. Examples of this are achondroplastic dwarfism (a disease that leads to retarded growth) and Huntington's chorea (a fatal disease that strikes later in life and leads to progressive physical and mental deterioration). Many other single-gene diseases are recessive and are expressed only when both chromosomes have the defect. Examples of these diseases are sickle-cell anemia (which strikes blacks and those of Greek descent and leads to the production of malformed red blood cells), cystic fibrosis (which strikes families of northern European descent and leads to a malfunction of respiratory and digestive systems), and Tay-Sachs disease (a degenerative disorder of the nervous system that strikes Jewish families of eastern European descent).

A third group of single-gene diseases embraces the sex-linked (also called X-linked) diseases. Because the Y chromosome in men is much shorter than the X chromosome it pairs up with, many genes on the X chromosome are absent on the homologous Y chromosome. Therefore men will show a higher incidence of genetic diseases such as hemophilia, Duchenne muscular dystrophy, and color blindness, which result from defective genes in the X chromosome. Even though these are recessive, males do not have another homologous gene on their Y chromosome that could contain a normal gene to mask it.

Another major cause of genetic disease is *chromosomal abnormalities*. Some diseases result from an additional chromosome. Down's syndrome (or mongolism) is caused by trisomy-21 (three chromosomes at chromosome 21). Klinefelter's syndrome results from the addition of an extra X chromosome (they are XXY). Other genetic defects result from the duplication, deletion, or rearrangement (called translocation) of a gene sequence.

The implications of the genetics of life are quite profound. Much of what we are is either determined or else conditioned by our genetics. As we can see, slight genetic errors can have profound effects. Thus one important goal of genetic engineering should be to find ways of treating and eventually curing these diseases.

THE SANCTITY OF LIFE

Another important foundation that should be built before discussing specific genetic issues is the sanctity of human life. Many genetic techniques impinge on one's view of the sanctity of life.

For centuries Western culture in general and Christians in particular have held to a view of the sanctity of human life. In our society today this view is beginning to erode into a quality-of-life standard. Where once we saw the disabled, the retarded, and the unborn as having a special place in God's world, now we have moved into a position of judging only the quality of human life. No longer is life as such seen as sacred and worthy to be saved. Now it is seen as something to be judged and evaluated. If we arbitrarily feel that life is not worth living (that it will not be a quality life), then it is advisable to terminate it.

Our discussion of these genetic techniques must begin with a recognition by us as Christians that human life is sacred because we are created in the image of God (Gen. 1:27; 5:1–2). We must not place an arbitrary standard of quality above God's absolute standard of human value and worth. We live in a fallen world and have to make difficult choices about children born with various genetic defects but we make those choices from an objective, absolute standard of human worth.

One important implication of the sanctity of human life is that *human beings are distinct from animals* because we are created in the image of God (Gen. 1:27; 5:1–2). This distinction is extremely important in a consideration of these genetic techniques. For example, artificial insemination is used to increase the genetic fitness of cattle. By using sperm and eggs from the fittest bulls and cows, cattlemen can implant fertilized embryos from the best stock in weaker strains of cattle and so improve the fitness of the herd.

If there is no distinction between people and animals,

then using that technique on humans would not create any ethical problem. But because there is a distinction, we cannot automatically transfer technology from animals to people. There are grave concerns over implementing a genetic breeding program on human beings using artificial insemination and embryo transfer.

A second implication of the sanctity of life is that *care for human life must include life in the womb*. Many genetic techniques threaten the life of the fetus, and so the moral status of the fetus is at the center of many of these discussions. If the human fetus is also a person, this fact will affect our conclusions concerning practices that are life-threatening to the fetus.

A biblical view of the fetus is that it has personality and value. In Psalm 139 in the Old Testament, David refers to his unborn state with personal pronouns. The focus of the passage is on him as a person and not some piece of protoplasm that would become David. There is a continuity between prenatal and postnatal life. In the womb David was already a person under the care of God.

In a number of passages in the Old Testament, a prophet is said to have been called by God when still unborn. For example, we read, "Before you were born I set you apart" (Jer. 1:15), and "the Lord ... formed me in the womb to be his servant" (Isa. 49:5). The Bible also makes little distinction between an infant and a fetus. The same word for "infant" is used in 1 Samuel 15:3 and Job 3:16, even though the latter passage refers to an unborn child.

In reflecting over the sin in his heart, David laments, "Surely I have been a sinner from birth, sinful from the time my mother conceived me" (Ps. 51:5). His sin nature was already present in him as a fetus. He was a personal, spiritual being while in the womb and already cast in the image of God. Old Testament scholar Bruce Waltke, after doing careful exegesis of various passages in the Old Testament (Genesis, Job, Psalms), came to "the inescapable conclusion that the image of God is already present in the fetus."[2]

In the New Testament the most important passage is the

first chapter of Luke. First, we see that although John the Baptist was in his mother's womb, he is referred to as a "baby" and is said to have "leaped" in the womb of Elizabeth (v. 41). Again, we see the continuity between life outside the womb and inside the womb. There are many passages in the Bible that refer to a fetus as though it is an infant. Second, the historic teaching of the Incarnation argues that Jesus was divine from the moment of conception. Graham Scott argues on the basis of the biblical teaching of the Incarnation in Luke 1:26–56 that human life begins at conception.[3]

GENETIC ENGINEERING

A variety of different genetic tools have been developed. The first is *genetic counseling*. Before a couple decides to have children, they may seek a genetic counselor. Each of us carry between three and eight genetic defects that might be passed on to our children. By checking the family medical histories and taking blood samples (for chromosome counts and tests for recessive traits), a counselor can make a fairly accurate prediction about the possibility of a couple's having a child with a genetic disease.

Most couples today, however, do not seek genetic counsel in order to decide whether to have a child but rather to decide if they should abort a child who is already conceived. Since the mother is already pregnant, the concern is not whether to prevent a pregnancy but whether to abort the unborn child. Many standard tests are given to assess the possibility of defects in the fetus.

Major deformities can be discovered by several new techniques. An important technique is *ultrasound*, which uses a type of sonar to determine the size and shape of the fetus. An ultrasound transducer is placed on the mother's abdomen, and sound waves are sent through the amniotic sac. The sonar waves are then picked up and translated into a TV screen picture that provides important information about the size and shape of the fetus.

Another important tool is *laproscopy*. A flexible fiberoptic rod is inserted by the doctor through a small incision in the mother's abdomen. This tool allows the doctor to probe into the abdominal cavity and determine if the Fallopian tubes are defective or if the pregnancy is proceeding normally. This instrument even allows the doctor to take a blood sample from the fetus.

An important diagnostic tool is *amniocentesis*. A doctor can insert a four-inch needle into a pregnant woman's anesthetized abdomen and withdraw up to an ounce of amniotic fluid. As the fetus grows, cells are shed from the skin, and these can be collected from the fluid and used to discover the sex and genetic make-up of the fetus.

Amniocentesis has become a very important test, and its use is now standard procedure in most hospitals. The greatest concern over the procedure is that its primary focus is on providing information so that a pregnancy can be terminated if necessary. While amniocentesis gives important diagnostic information that can save the life of the fetus, it also provides information that may threaten the baby's life.

Amniocentesis is mostly used to determine if the fetus has any genetic defects. The majority of these defects cannot be cured at present; so the tacit assumption is made that if defects are found, the couple will opt for an abortion.

There are few other major reasons for providing this information, especially in the earlier stages of pregnancy. If the genetic disease cannot be cured, then why tell the prospective parents while the child is yet unborn? A prior assumption of abortion precedes most tests using amniocentesis. In many cases a doctor or clinic will not do an amniocentesis unless a couple will consider having an abortion.

The pressure this places on a couple is very great. Providing this information often "forces" couples into decisions they might not have otherwise considered. There are cases of couples who have had an amniocentesis with the assumption they would not abort but later changed their minds on receiving the information.

Of even greater concern is the general trend toward mandating an abortion when genetic defects are diagnosed through amniocentesis. In a case involving Tay-Sachs disease,

Court of Appeals Judge Bernard Jefferson ruled that not only the parents but possibly even the physician and laboratory could be sued for negligence in not having aborted the fetus.

In the area of reproductive biology, there have been many advances. The major goal of reproductive research has been to offer alternatives to the normal processes of reproduction. Nearly six million couples of childbearing age are infertile. Most estimates seem to indicate that infertility is a growing problem.

Dr. Ralph Dougherty of Florida State University has shown that the sperm count of American males has fallen 30 percent in the past fifty years. Due to such factors as environmental pollution, nearly a fourth of all men now have sperm counts low enough for them to be considered "functionally sterile."[4]

Female infertility is also a problem due in large part to the rise in promiscuous sex. Dr. Robert Francoeur, professor of human sexuality and embryology, notes, "Because more young women are having sex with a variety of partners and sustaining low-level gynecological infections that may go untreated and damage the reproductive system, as many as one in four women between the ages of 20 and 35 are now infertile."[5]

Artificial insemination is used as an alternative means of reproduction when male infertility is present. It has also been used more recently as an alternative to female infertility. Women have been impregnated with donor sperm so that couples can adopt children born to these surrogate mothers.

In-vitro fertilization is another means of reproduction when a female is infertile. In this method, conception takes place outside of the womb; hence the popular description "test-tube babies." A third development in this area is *artificial sex selection.* When this procedure is used, the sex of a child can be determined before conception.

In the area of genetic manipulation there are various methods that have been proposed for modifying organisms and controlling their genetic future.[6] The first is *eugenic engineering.* The focus here is on controlling the recombination of genes by directed control of conception through parental selection. This can be done in two ways. First, there is positive eugenics involving selection for particular traits. Second, there is negative eugenics involving an attempt to breed out certain undesired genes. Although some attempts at eugenics have been proposed for humans, most applications of this method are found in agriculture and animal husbandry through the use of selective breeding.

The second method is *euphenic engineering.* This is the attempt to modify gene action in order to regulate certain deleterious effects on genetic disorders. The focus here is on treating the genetic disease in order to minimize its ill effects. A common example of this is the treatment of juvenile diabetes. Diabetics suffer from an inability to manufacture a sufficient quantity of insulin in their pancreas and must have insulin injected into their bodies in order to compensate for this genetic defect.

A third method of genetic manipulation is *gene engineering.* The intent of gene engineering is to "fix" or remove the offending genes. This will probably be done some time in the future by using gene surgery to remove the defective genes from the chromosome and replace them with new ones. Advances in this field suggest that in the future this may be done by growing a virus with the necessary genes that will then be inserted into the defective chromosome. Advances in *recombinant DNA research* offer the promise of using gene surgery on humans in the future.

In the following chapters we will look at some of these various genetic techniques. Although many of them raise similar scientific, social, legal, ethical, and theological problems, each particular technique has a unique arrangement of problems. Each has its own set of benefits and dangers that must be analyzed. Our responsibility as Christians is to carefully think through these issues in our future genetic revolution.

Artificial
Reproduction

II
Artificial Insemination

The focal point of artificial reproduction is to cure infertility. One major cause of infertility in couples is male sterility. Sterility in men is often caused by accidents, chemotherapy, or environmental pollutants. Artificial insemination techniques are often used as a means to overcome problems of sterility.

Artificial insemination was first done with humans in 1785 in London by a doctor named John Hunter. Today there are two types of artificial insemination: using semen of the husband (AIH) and using semen of a donor (AID). As a medical procedure artificial insemination is a relatively simple process. Sperm is collected from the husband or donor through masturbation. Special buffer solutions and glycerol are then added to the sperm. After examination of the sperm it is stored frozen in liquid nitrogen until it can be thawed at room temperature and injected into the woman during her peak phase of ovulation.

ARTIFICIAL INSEMINATION BY HUSBAND (AIH)

Artificial insemination by the husband consists of collecting the husband's sperm and injecting it into his wife. Couples may seek this procedure for a number of reasons. First, the husband may be fertile but unable to participate in normal sexual relations. Second, the husband may have a low sperm count. His sperm can be collected periodically and its volume increased so that when it is injected into his wife, there is a greater probability of pregnancy. Occasionally, a doctor may mix the husband's sperm with a donor's seminal plasma. Since the donor's spermatoza can be separated from the plasma, the pregnancy would be from the husband and thus still be considered AIH.

Third, the husband may use AIH as a precautionary measure. He may be planning to have a vasectomy and still have some precaution for the future. AIH is also used by those about to undergo prostate surgery in order to assure the potentiality of future procreation.

Since there is much less controversy surrounding AIH than AID, we will deal with it in less detail. It is a relatively effective means by which a couple can produce a child who is genetically related to them. Although the sexual act is not natural, using artificial insemination with the husband does not destroy the personal and sexual aspects of the marriage bond and thus has been less criticized.

There are few legal concerns over the method, since the child is genetically related to the couple and any question of legitimacy of the child is therefore removed. Questions of paternity or inheritance are not a problem for children conceived by AIH.

There are few theological concerns as well. The major theological issue that has been raised is the question of masturbation. In particular the Roman Catholic church has objected to masturbation because it separates sex from the conjugal relationship and may lead to an enslaving habit.

To claim that masturbation in this case is sin is to remove it from its context. Although a couple does not experience the sex act together, the purpose is to provide a pregnancy and birth that they will experience together. AIH does not seem to inhibit sexual expression in the couple or damage the marriage bond; thus most Evangelicals have seen it as a legitimate intervention into a marriage.

ARTIFICIAL INSEMINATION BY DONOR (AID)

The scientific aspects of AID are similar to those of AIH except that a sperm donor is used instead of the husband. This singular exception has led to a number of questions. Most of these questions center around either the legal status of a child born by AID or its impact on the marriage bond.

About twenty thousand children are born each year through AID. Entire companies such as IDANT in New York and sperm banks around the country collect sperm and make it available to physicians. The sperm sample is usually obtained from undergraduate students or medical students who have had their genetic history checked very carefully. After some counseling with would-be donors, the physician looks for a sperm donation. Most doctors attempt to match the physical characteristics of the husband with those of the sperm donor.

Couples usually seek AID for one of three reasons. First, the husband might be carrying a genetic disease that he does not want to pass on to his child. Second, he may be sterile as a result of a disease or accident. Third, AID might be considered when there is some concern over an antibody reaction from the mother when, for example, the husband is Rh positive and the mother Rh negative. In most cases, however, injections of anti-Rh antibodies into the mother are sufficient to prevent damage to future Rh positive children.

SCIENTIFIC CONSIDERATIONS

Most of the concern with AID is not scientific, but a few points should be mentioned. One is the fact that sperm donors remain anonymous. Although there may be a number of good reasons for keeping the donor anonymous, one unfortunate fact is that it increases the possibility of accidental incest.

Researchers at the University of Wisconsin have found that an average sperm-bank donor is used for up to six pregnancies and some for as many as fifty pregnancies.[1] Since the identities of donors are kept secret (often even from the physician), the potential for inbreeding between the resulting half brothers and half sisters later in life is very great. This is a significant problem in small communities with less mobile populations. In a short period of time a large percentage of the population might become genetically related (half siblings or cousins) just by the overuse of a particular sperm donor.

Another problem is the accidental transmission of unknown genetic traits. If a child later contracts a genetic disease, the couple bears the burden even though the malady was transmitted by the donor and not by the husband.

Studies have shown that the amount of genetic screening done on sperm donors is far from exhaustive. A survey of 471 physicians who were members of the American Fertility Society showed that 80 percent of them had performed AID. Most of the doctors said they took only oral histories from the donors and only 29 percent even performed a blood test for communicable diseases.[2]

It could be argued that this is the risk a couple takes by using AID and they should be aware of that possibility. Although we are all playing a game of genetic roulette, there is no reason that we should not provide greater scrutiny over sperm donations. There are many reports of various genetic diseases that have been transmitted through AID,[3] and some level of protection needs to be assured those who resort to this procedure.

LEGAL CONSIDERATIONS

The use of AID is presently in a legal tangle. Most state laws concerning parent-child relations were not written with artificial insemination in mind. The legal status of the couple in relationship to the child produced by AID is still uncertain.

The major focus is on the legitimacy of the child. Since the child born of AID is not genetically related to the father, there is the potential that the child will be declared illegitimate. The only way to protect the child's legal relationship to his father is to presume that the child is legitimate in the face of obvious evidence to the contrary.[4] Presently in the United

States only fifteen states have enacted legislation to protect the rights of an AID child.[5]

There are three important reasons why a child conceived through AID must legally be declared the child of the couple. First, there must be *certainty of child support* in the event of desertion or divorce. A husband could leave the family and then claim that he does not owe any support to the AID child, who was not his natural child. Unless the legal relationship is established, the issue of child support will always be in question.

Second, there must be *clear lines of legal inheritance.* In the event of a death in the family, it could be argued that the AID child is not a lawful heir. This uncertainty of inheritance must be cleared up through the enactment of state laws. Only fifteen states have statutes that grant legal inheritance rights to AID children. The other thirty-five must follow suit. If this is not done, a further danger may emerge. Since many of the sperm donors are former medical students whose level of income has increased greatly, the AID child may be able to sue the sperm donor for child support or inheritance.

Third, there must be legal precedents to establish that *AID does not constitute adultery.* In occasional cases courts have ruled that AID constituted sufficient grounds for divorce due to adultery.[6] If AID is going to be allowed in our society, then clearer guidelines must be established. State legislatures cannot shirk their responsibility in the light of the twenty thousand AID births a year that are in a significant legal tangle. These AID births are no longer a minor problem but present a major agenda that must be confronted.

If a child is born through AID, there should be a relatively simple procedure to establish the legitimacy of the child. In order to protect the anonymity of the sperm donor and the privacy of the family, it would be best not to make this a formal courtroom procedure. This would then remove the questions of inheritance and child support, and would significantly weaken the case for AID as the basis for divorce. State courts should refuse to hear cases that cite AID as grounds for divorce, and precedents should be established.

SOCIAL CONSIDERATIONS

Although the legal considerations of AID are uncertain, the social implications are quite disturbing. If it is to be allowed in our society (this will be addressed more fully in the ethical considerations section) as an alternative means of procreation, then stricter guidelines must be implemented concerning its use. The unrestricted use of AID will lead to a number of serious social problems.

The first problem is that AID *will increase the number of single-parent families.* About 9 percent of those seeking AID in this country are single women.[7] Therefore, about fifteen hundred children are being born into single-parent relationships each year.

It is sad that already so many children live in broken homes with a single parent, but it should be of even greater concern that so many children are being intentionally brought into a relationship without any father figure. At least the children of a divorce know that their father exists and that they are living in an abnormal situation. Children conceived by a single parent are placed in an extremely unnatural environment. The effect on their mental stability and sexuality must be questioned.

In some even more bizarre cases, AID is being used to bring children into very unfortunate situations. For example, dozens of lesbians in England have used AID to give them children for their "lesbian marriages."[8] The number of cases in this country is unknown, but it is likely as high and destined to go higher. In an even more unusual case, a man who had a sex change operation to become a woman decided to become a "mother." The transsexual used artificial insemination from her husband to impregnate his (now her) sister and then adopted the child.[9]

A second problem is that AID can adversely *affect the marriage relationship.* The psychological impact of AID can be quite profound in many marriages. For example, the need to resort to AID often comes as a blow to the husband's masculinity.[10] His inability to produce children can develop into a deep feeling of failure or inferiority within a marriage relationship.

The psychological trauma is further heightened by the fact that much of the procedure is kept secret. The couple rarely know the identity of the donor and often do not make public the fact that they have used AID. This secrecy often gives the procedure an illicit aura and begins to reinforce feelings of guilt.

A couple with an AID child may have difficulty explaining "who the child looks like," since they often wish to keep the procedure secret. A couple with an adopted child has less difficulty with these problems than a couple with an AID child who, many may suspect, was conceived through some extramarital affair.

One New York psychotherapist found that a number of her patients were AID mothers struggling through feelings of guilt and fear. She found that the couples were often reluctant to tell their children how they were conceived. Over a period of time the secret tended to assume "undue proportion and power within the family" and became "an existential fact that remains unspoken yet controls and colors the lives of the people involved."[11] Although these problems usually do not lead to divorce (the divorce rate among AID couples is much lower than the national average), they can cause additional tensions that a couple should be aware of.

A third problem is that AID has contributed to *the rise of surrogate mothers.* The group that has attracted the most media attention in this regard is Surrogate Parenting Associates (SPA) in Kentucky. Dr. Richard Levin, a Louisville fertility specialist, is president of the association and is aided by a young attorney named Katie Marie Brophy. They have developed a computerized system of applicants, and, for a fee, an applicant can agree to be the surrogate mother for a childless couple. Surrogates are chosen from applications received from newspaper ads, and the women are cataloged according to their physical and mental characteristics and ethnic and religious backgrounds.[12]

The first such paid surrogate mother was Elizabeth Kane (a pseudonym), whose baby was born in November, 1980, and who was under legal contract with Mr. and Mrs. Ralph

Ransdale. The couple adopted the child in April, 1981, using a procedure similar to that followed for stepparent adoptions.[13]

Another person in this field is Noel Keane, a Dearborn, Michigan, attorney, who arranges voluntary matches and has about twenty-five couples seeking surrogates and has five pregnancies under way. He is concerned with Levin's approach of payment for services and fears that a woman who was contracted to carry a child for $10,000 may decide she now wants $50,000 for the services.[14] Already one woman has advertised her services as a surrogate mother in the classified ads. She is asking a fee of $15,000 in order to pay her way through nursing school.[15] It is certainly possible that this growing field of surrogate motherhood may develop as many problems as the current practice of selling black-market babies.

The legal problems this new development brings are very great. The contract used by SPA covers almost every possible contingency. The surrogate mother agrees to be artificially inseminated by the prospective father, carry the child full term, and transfer custody of the child to the donor couple. She also agrees not to smoke or take medication without Levin's consent.

The contract's greatest weakness is that it can't really be enforced. There has been some question as to whether charging a fee constitutes the sale of a baby and thus violates state law. Sanford N. Katz, chairman of the American Bar Association's Family Law Section, say, "I know of no court that would be sympathetic to a contract involving the selling of babies."[16] Besides this technicality there are many very sticky problems.

If the man or woman dies before the child is born, will the surviving member be able to adopt the child? What if the surrogate mother decides to keep the child? Who is the legal mother? If the surrogate mother decides to abort the child, can the donor couple sue for damages? Who is responsible for a defective child: the prospective parents, the surrogate mother, or the physician performing the AID?

These questions do not now have definite legal answers, and only future court cases and state laws will help to resolve them. At present there are two court cases that may slow surrogate parenting. One case involves a court attempt by the Kentucky attorney general to issue a permanent injunction against Levin's group for violating at least three state laws (adoption and solicitation codes).[17]

A second case involved a surrogate mother, Denise Lucy Thrane of Arcadia, California, who wanted to keep the child she bore. Mrs. Thrane contracted with Mr. and Mrs. James Noyes of New York through Noel Keane to be artificially inseminated and carry the child. Although she was divorced and had three children of her own, Mrs. Thrane decided to keep the child. The case went to court in Los Angeles and was about to be heard in the Los Angeles Superior Court. However, the sperm donor, James Noyes, withdrew his suit because his wife was a transsexual and he feared that the publicity would not allow his child to lead a normal life.[18]

There are apparently only two other cases in the United States involving surrogate mothers. One case is pending in Michigan over whether money may be paid for surrogate services. Another in Kentucky is being fought on the question of prohibiting a surrogate mother from giving up her rights to the child for a fee.[19] In England there was a case involving a man who hired a prostitute to carry his child but lost custody when she decided to keep the baby.[20]

A fourth problem is that AID has led to *the development of human eugenics programs.* Although the hope of human breeding was crushed strongly after World War II, the idea is not completely dead. For example, there is the recent announcement that a sperm bank for Nobel Prize winners has been established in southern California.

Robert Graham, an eccentric multimillionaire, stores sperm from Nobel prizewinners and other high-IQ participants in an underground vault cooled by liquid nitrogen. At his Repository for Germinal Choice he has already begun fertilizing women from Mensa (an organization for those with high IQs) and hopes to impregnate hundreds of other intelligent women in order to produce children of superior intelligence and high potential. Graham argues,

The principles of this may not be popular, but they are sound. We're trying to take advantage of the possibilities of genetics. We are hoping for a few more creative, intelligent people who otherwise might not be born.[21]

Although Graham has said that he is not trying to create a super race, many critics cannot help but see the similarity between his program and Nazi Germany's attempt to develop a "master race." Geneticist David Baltimore feels that the scheme is "dangerously elitist," and many Nobel laureates have declined to donate sperm.[22]

ETHICAL CONSIDERATIONS

Many of these programs create significant ethical problems. Therefore we should consider some of the ethics of the practice of AID in our society today.

The first ethical issue is *the slippery-slope argument.* We seem to be slipping down the slope of moral choices into immoral consequences. It is one thing to consider AID as a last-resort option for couples suffering from male infertility. It is quite another to allow it to produce children for single parents, lesbian parents, and transsexuals.

Bringing children into an abnormal relationship should raise ethical and theological concerns among Christians, who value the importance of the home and family. We may regret that death, desertion, and divorce create one-parent homes, but these situations do not justify bringing other children into similar situations through AID.

How will a child growing up in such an environment develop a proper sense of God's plan for the family when one partner is not present and will never be present? When asked how she would explain her decision to have her child through AID, one woman said, "I will tell him later that I was very much in love with his father who died in an airplane crash."[23] A lie will be at the very foundation of her relationship with her AID child.

The implications of surrogate motherhood are also distressing. While it may be a compassionate act to carry another's child, the trajectory points toward a world of genetic confusion. Because of the uncertainty of whether a surrogate mother will carry the child to term and then relinquish it, the next logical step will be to retrieve the embryo from the surrogate womb and implant it in the adoptive mother.

Now all sorts of possibilities begin to emerge. With the addition of in-vitro fertilization, it would be possible to mix sperm and egg from *any* two people and put the fertilized embryo in *any* womb and have *any* set of parents (or single individuals) adopt the infant.

This scenario raises the specter of a world of commercialized baby markets. There is already a considerable traffic in babies in this country. Prices have already gone as high as $50,000 for custom-created babies in Los Angeles. A few years ago actor Richard Burton gave an English magazine a handwritten ad for a woman under thirty-eight who would bear him a child on a contingency-fee basis: $25,000 for a girl and $50,000 for a boy.[24]

With artificial insemination, mix-and-match children could be offered with genetic specifications and prenatal environment guaranteed. One childless couple, who paid $50,000, selected both the male inseminator and the female bearer from a collection of photographs of attractive single men and women.[25]

We now have the bizarre situation in which unborn babies reared by surrogate mothers are being traded like merchandise and fought over in court battles. This scenario will only grow more bizarre if we do not guard against it.

Future developments will also be determined by what we do with the proliferation of sperm banks in this country. Present estimates suggest that there are well over one hundred thousand sperm samples available. How long will it be before we see celebrity sperm being advertised? Will the going rate for the sperm of some popular star be higher than the national average? Will sperm banks be established for Pulitzer prizewinners as well as for Nobel prizewinners?

The underlying assumption seems to be that our increasing knowledge of genetics will solve all our problems. Yet the history of the world has shown us that we suffer much more at the hands of moral defectives than we do at the hands of genetic defectives. Genetic purity is not the solution so much as is moral purity.

A second ethical concern is *the place of reproductive technology in our society*. In many cases a technological "quick fix" is being offered for social problems. For example, there seems to be a direct tie between abortion and surrogate motherhood. Many couples might be quite willing to adopt a child rather than go to the effort of finding a surrogate to carry their child. But because there are over a million abortions a year it is difficult to adopt an infant. As a result artificial reproductive technology has been developed in order to satisfy the demand for children. Surrogate mothers do not solve the underlying social problem but only treat the symptoms.

It is also very interesting to hear surrogate mothers encouraging others to "share their bodies." The phrase is so very close to the proabortion line: "women controlling their own bodies." Somewhere in all of the rhetoric, the individuality of the child is lost. Everything begins to be reduced to the physical (sperm and egg), the possessive (my body, my child), and the material (womb for rent). The lowering of human value seems inevitable. As one article insightfully asked of Surrogate Parenting Associates in Kentucky, "In a state where horse-breeding is a major industry, can brood-mare mothers be far behind?"[26]

THEOLOGICAL CONSIDERATIONS

One of the most important theological issues in AID is *the biblical view of sexual relations*. Artificial reproductive technique is obviously a deviation from God's intended purpose for sex in marriages, but any rejection of it must be based on more than just a feeling that it is unnatural. AID raises two fundamental questions. First, is AID an act of adultery? Some commentators have felt that AID shares a great deal of similarity with adultery in that a wife becomes pregnant from someone else's sperm.

There are, however, significant differences between AID and adultery. Adultery involves sexual infidelity by a sexual relationship between a married person and someone not his or her spouse. AID involves merely the transfer of gametes between consenting adults. First, there is no sexual contact between the wife and the sperm donor. Second, there is the mutual consent of the husband and wife. These differences seem to invalidate the comparison of AID with an illicit act of adultery.

One of the major factors in adultery is the *attitude*. Jesus said that "anyone who looks at a woman lustfully has already committed adultery with her in his heart" (Matt. 5:28). There is no such attitude discernible in AID. Another factor is *action*. Paul talks about becoming "one body" with a prostitute (1 Cor. 6:12–16); such a union is not found in AID. Therefore it is not appropriate to call AID a form of adultery.

A second question is whether AID damages the sexual act. The Bible discusses two important aspects of sexuality in the bonds of marriage. First, there is the unitive aspect (Gen. 2:24) of becoming "one flesh." Second, there is the procreative aspect (Gen. 1:28) of being fruitful and multiplying. AID separates these two functions and reduces procreation to a biological act.

It should be noted that in a typical AID case there is little concern over this problem. Both the husband and the wife consent to the medical procedure, and the third party does not invade or alter the one-flesh unity of the couple. Although masturbation is involved, the context in AID is similar to that in AIH, which is a procedure generally accepted by most ethicists. AID differs from AIH in that the child produced is only partially a product of the couple.

There is some biblical support for the possibility that God may allow artificial insemination to be used in special situations. For example, Deuteronomy 25:5–10 records the provision of the levirate marriage of the kinsman-redeemer, whose responsibility was to impregnate his deceased brother's wife if there was no heir. There is an obvious similarity between this example and artificial insemination.

There are also some important differences between the two that should be noted. First, this levirate marriage was instituted within the Old Testament theocracy. Today God is not calling for an unmarried brother to take his dead brother's childless wife. The levirate marriage was instituted to ensure a line of descendants.

Second, the unitive and procreative aspects of marriage remained intact. This unusual arrangement still guaranteed that unity, sexuality, and procreation were still joined together in one act. Therefore, while it may be tempting to justify AID by comparing it to the levirate marriage, there is no one-to-one relationship. The principle, however, is important. God allowed another party (the brother) to provide a child. The Old Testament provision was to ensure descendants for a family line. AID could be seen as the modern counterpart to provide a child for a married couple.

To allow AID as a possibility does not argue for its complete acceptance. Rather, we may reason that since we live in a fallen world (Gen. 3) where some couples are infertile, AID may be permitted in certain cases to bring children into a loving environment. It can be a method of last resort for couples who have sought other remedies and have no other hope. The principle of the levirate marriage is that God may allow the unusual in order to continue His blessing (Gen. 17:7–9).

A second important theological issue is *the biblical view of parenthood.* God ordained marriage to be the union of two people who would give birth to a child who is genetically related to them. While there are obvious exceptions to that rule (e.g., adoptions), the divine ideal should be the standard used to judge AID. While AID may be allowable in particular cases of last resort to a married couple, this does not sanction its use outside the bounds of marriage. Producing children for single parents and lesbians or making surrogate arrangements goes far beyond the bounds of a biblical view of parenthood.

The one Old Testament example that might seem to sanction surrogate motherhood is found in Genesis 16. When Sarah could not bear a child for Abraham, she said to him, "Please go in to my maid; perhaps I shall obtain children through her" (Gen. 16:2 NASB).

There are a number of reasons for rejecting this as an

appropriate biblical precedent. First, there is no indication of God's approval of the act. In fact, it seems that Sarah's suggestion merely clouded God's lesson for them. He wanted them to be dependent on His provision for a child, but they attempted to circumvent His desires. Second, the action results from a polygamous relationship allowed by a culture that does not exist today. Also, since there was no separation of the procreative and unitive aspects of sexual intercourse, the biblical incident is not really analogous to what occurs in AID.

Surrogate mothers significantly blur the distinctions of a true view of motherhood. Artificial insemination of a surrogate dehumanizes the aspect of prenatal care for the infant and opens up a realm of commercialization of fetal rights. This comes into sharper focus when we begin to consider what might occur if it were found that the surrogate mother was carrying a fetus with a genetic defect. The surrogate might want to abort, but the prospective parents might file a court order to prevent the destruction of the "property." Or it is possible that the prospective parents might want to abort, but the surrogate might not want to undergo an operation that could threaten "her body." It is also likely that neither party would want the child; so the government would be forced to declare the infant abandoned and so to be regarded as a ward of the state.

Somewhere in the process a child has ceased to be a gift of God (cf. Ps. 127:3) and has become a commercial item. If you buy a product and it is defective, you return it for your money. If a child is born with a defect, the Christian view of the sanctity of life calls for us to minister compassionately to the one in need. Somehow, children are becoming less human and more commercial as a biblical view of parenthood breaks down.

In conclusion, we can sanction AIH as a means of providing children for couples suffering from low sperm counts or sexual dysfunction. AID may be allowed in principle for married couples as a last-resort method. There are, however, particular dangers and concerns that should be discussed with the couple. There is a similarity between AID and pregnancy from an extramarital affair that may affect the stability of the marriage bond. For this reason there should be counseling sessions with the couple in order to determine if this similarity would affect their marriage relationship. They should be prepared for the psychological stress that such a decision might entail. The use of AID for extramarital pregnancies should not be allowed especially in view of the nature of our society and the unethical world view that is prevalent today.

III
Artificial
Sex Selection

Artificial insemination has spawned a separate question concerning artificial sex selection. Many are asking whether parents should be allowed to select the sex of their child by using new sperm separation techniques that utilize insemination as the means of sperm delivery.

The discussion of the topic has become significantly muddled because of the fact that couples occasionally have used abortion as a means of sex selection. By using amniocentesis a doctor can determine the sex of a fetus, and the mother is then free to use that information for sex selection. Although most of this chapter will focus on preconception sex selection, it is important to first consider the question of abortion and sex selection.

ABORTION AND SEX SELECTION

In amniocentesis a doctor withdraws fluid from around the fetus and uses that fluid to diagnose a number of genetic diseases. Some of the fetal cells in the fluid can be grown in a laboratory mixture, and then within a few weeks a genetic map can be made of the fetus inside the mother's womb. This map provides information about the genetic structure and sex of the fetus.

This additional information is potentially dangerous. While some couples may appreciate knowing the sex of their unborn child in order to plan the nursery, others may seek this information so they can abort a fetus of the "wrong sex." A number of cases of abortion for this reason have been documented in this country,[1] but there is no way of knowing how many abortions have been sought for this purpose.

It is curious that so many people who otherwise support the practice of abortion on demand react against this procedure. Somehow an abortion for personal convenience is less repugnant than one for sex selection. Abortions for genetic defects are encouraged, while abortions for sex selection are discouraged.

This is where the abortion movement becomes most inconsistent. The same people who sanction abortions for women carrying a fetus with an extra chromosome (which will develop Down's syndrome) react against an abortion for a fetus with an extra X chromosome (a girl). Quantitatively, there is little difference between the two abortions, since both involve one unwanted chromosome. Once society has sanctioned abortion for genetic reasons, it is nearly impossible to argue that only some genetic traits may be a just cause for therapeutic abortions. The justification for a Down's syndrome abortion today will be the justification for "wrong sex" abortions tomorrow and for "wrong hair color" abortions in the not too distant future.

The Bible establishes the basis for human dignity. The fetus as a person must not be destroyed simply because he or she does not measure up to parental criteria. Since the fetus shares our humanity, Christians should not allow the use of abortion as a means of sex selection.

The question about other modes of sex selection is not so clear. In these other methods the sex is selected *before* conception. With the abortion method the sex is selected *after* conception. The former involves directing the conception, while the latter involves eliminating unwanted conceptions.

SEX-SELECTION PROCEDURES

Sex selection can be performed both naturally and artificially. Although the focus of this chapter will be on the artificial methods, since they are more accurate as well as more controversial, a brief mention should be made of the natural methods.

Around the world there are wives' tales and folklore concerning sex selection. Aristotle advised a woman to lie on her left side to conceive a daughter and on her right side for a son. Some people say that a man should hang his pants on the right bedpost before sexual intercourse in order to have a son and on the left for a daughter.

Various diets also have been suggested for sex selection. One authority suggests eating foods rich in fats and sugars for a girl and eating salty, spicy foods for a boy. Unfortunately, there is little evidence that such diets will produce anything but indigestion.

David Rorvick in 1970 wrote a book entitled *Your Baby's Sex: Now You Can Choose.*[2] It focused on certain physical indications, such as those used in the rhythm method of birth control, to increase the likelihood of having a child of a particular sex. Although fairly popular and gadget-free, most scientists consider it ineffective.

The techniques that promise much greater certainty involve artificial intervention. The sex of the child is determined by the sperm of the male. Sperm with Y chromosomes will produce boys, while sperm with X chromosomes will produce girls. This fact has provided researchers with a means by which to direct the conception of a child of the desired sex.

About five different approaches to sperm separation can be used to direct the sex of a child. Two of these show particular promise. One is to immobilize the sperm cells in a refrigeration unit. The sperm cells that could produce a girl (X sperm) tend to be a little heavier and thus settle to the bottom of the liquid mixture faster than Y sperm.

Another technique is to allow the sperm to swim through a dense, viscous fluid (bovine albumin). The Y sperm cells move more readily through this fluid than do the X sperm cells. This procedure can provide up to 85 percent male sperm.[3] If these techniques are used together, the probability for success approaches 100 percent. The desired sperm can then be drawn off of the mixtures and used in a standard procedure of artificial insemination.

Dr. Ronald Ericsson, a reproductive physiologist in California, has developed one of these sex-selection techniques and has been doing animal studies on his ranch in Wyoming to determine its effectiveness. Clinical application of Ericsson's techniques has been done at Chicago's Michael Reese Hospital under the direction of Dr. Paul Dmowski. His primary focus is on helping couples to increase the likelihood of having a boy, and he has been relatively successful. Researchers assume that these techniques change the success ratio to 70:30, but there are still too few births to establish the actual probability of success.[4]

Some scientists predict that pharmacological intervention may be used in the future to determine the sex of a particular child. New drugs could act in a way similar to birth control except that they would create a change in body chemistry resulting in an increase in production and survival of particular sex-determining sperm cells. Australian biologist Charles Birch foresees the day in which sex determination pills will be produced—pink for girls and blue for boys.[5]

POTENTIAL BENEFITS OF SEX SELECTION

Three major benefits are often cited for the implementation of artificial sex-selection procedures. First, it is argued that sex selection will aid in *family planning*. This would allow a couple to plan their family much more exactly. Dr. Paul Ehrlich at Stanford University believes that many couples would limit their families to two children if they could be assured of having a boy and girl.[6]

This method of family planning would especially be of interest in many underdeveloped countries that have very high population growth rates, which are often linked to the intense desire of parents to have sons. Sons assure social and cultural advantages, and third-world families often produce more children than they can adequately manage as they attempt to have sons.

Government workers in these countries have expressed

interest in sex-selection techniques that provide a means by which to interest their citizens in birth control. It is argued that if the birth of sons can be assured, the population growth rates of the countries will be reduced significantly.

A second benefit often cited for sex selection is *the prevention of child neglect and abuse.* It is argued that this procedure would prevent a child from being shunned because he or she was of the "wrong sex." In our society many children suffer from neglect and even abuse because they did not live up to their parents' expectations. It is not uncommon for a girl to grow up and feel the resentment of parents who wanted a boy instead.

It is naïve, however, to believe that sex selection will solve this social problem. The practice of abortion has not reduced child abuse in the United States. The argument that abortion would reduce the number of unwanted children and thus reduce the number of child-abuse cases has been proved false. Many of the abused children in our society were so-called "wanted" children.

There is some indication that the implementation of this procedure would create a climate that is even more hostile to at least certain children. With the current fifty-fifty average, a child of the "wrong sex" is seen as an unavoidable situation. In a culture in which sex can be controlled more exactly, a child who is still of the "wrong sex" may become even more neglected and abused and perhaps ostracized as an "accident."

A third benefit would be the possibility of *reducing genetic disease.* A number of genetic diseases are sex-linked and the power to have a girl, for instance, would eliminate the possibility of having a child with hemophilia.

SOCIAL CONSIDERATIONS

In determining the social implications of sex-selection procedures, the first important question to be answered is, Would couples choose the sex of their child if they had the chance? Most surveys indicate that they would if the procedure was acceptable and effective. In the 1970 Fertility Study directed by Norman B. Ryder and Charles F. Westoff at Princeton University, it was found that 39 percent of the six thousand women of childbearing age would choose the sex of their child if they had the chance.[7] A follow-up study four years later by Charles F. Westoff and Ronald R. Rindfuss showed that 47 percent of married couples would choose the sex of their child, and there were indications that the percentages would go even higher when a procedure became available and generally acceptable.[8]

Of those couples who expressed a preference for a particular sex for their child, it was found that about 90 percent wanted the firstborn to be a boy. If they could have only one child, 72 percent wanted a boy. If these two figures are combined, 75 percent of the firstborns and about 64 percent of the children in single-child families would be male.

This leads to a rather disturbing irony. One of the goals of artificial reproduction is to give women greater opportunity for involvement and decision making and thus greater self-esteem. However, this liberation of human procreation may have unintended consequences. The National Right to Life Committee's Dr. Mildred Jefferson has warned that this new procedure might "liberate women out of this world."[9]

The actual impact would be difficult to determine. Surveys of individuals give only a rough guideline of actual practice. Sex selection may well be influenced by such unmeasurable variables as social pressure and personal taste; so it would be difficult to determine how many couples would use these procedures. In the Western world the average family has about two children; so it is likely that sex-selection procedures might be used to produce one girl and one boy. But there may be enough of an influence in one-child families and larger families for the present fifty-fifty ratio to be appreciably changed.

Some social scientists have feared that this alteration would significantly change the social climate. These fears are somewhat exaggerated. A higher percentage of males may create a more rugged, hostile society as one may find in Alaska (where men outnumber women), but as much of this phenomenon can be attributed to the environment as to the greater proportion of males. Homosexuality may rise, but this would more likely be due to social factors that are already at work rather than to the implementation of sex-selection procedures.

Columbia University sociologist Amitai Etzioni has noted that men are much more likely to become criminals than are women, while fewer of them attend church or are involved in the moral instruction of the children. He therefore concludes that such sex-selection procedures will have some social effects that may return us to a more violent, frontier type of society.[10]

Some of these predictions may result, but it is also quite likely that any imbalance in the present sex ratios would be short-lived and relatively inconsequential. Unless sex-selection techniques become nearly universal, the sex ratios would not change significantly in Western-world countries. Further, it is generally believed that even if a major imbalance were to take place, a wave of male births would be quickly followed by compensating female births so that the oscillations would soon dampen out.[11] Also, the current movements for greater feminine equality should have some effect on parental perspectives and perhaps lessen the desire to have boys rather than girls.

ETHICAL CONSIDERATIONS

An important ethical question is whether parents have the right to determine the sex of their child. A broader question is whether society at large has the right to decide the sex of children to be born into it.

Part of the answer to this question is theological and will be considered in the following section, but most of this question must be answered with reference to our present social and cultural context. At a time when more and more rights are being claimed, it is quite likely that parents will begin to demand the right to determine the sex of their children. It is in this demand for rights that the principal danger with this new technology lies. The right of sex selection will most likely lead to a call for the right of genetic specification.

In a sinful world the slide down the slippery slope from moral actions to immoral choices and consequences happens quite easily. Somewhere we must cease our attempts to fulfill particular desires and concentrate on our responsibilities to our children and society. At some point on the continuum children cease to be a gift from God (cf. Ps. 127:3) and begin to be a parental plaything.

At a time when women are already claiming authority over the fetus they carry, we should question the motives of parents who claim the right to control the sex of their children. Sex selection may be *allowed* in certain instances, but it certainly does not qualify as a *right*.

Another ethical concern is sexual stereotyping. The present chauvinistic condition of our world may be a sufficient reason for limiting the use of sex-selection procedures. Sex selection could be the ultimate form of sex discrimination. Most families would want a boy first. Being born first seems to give a child an advantage in our society. First-born children tend to be more aggressive and achievement oriented than later children. The "boy-first" syndrome would most likely perpetuate existing stereotypes.

Much of our identity is conditioned by our sexuality. By choosing a person's sex, one also chooses part of that individual's personality. A society that already places maleness on a higher scale than femaleness is probably not ready "for the responsibility of assuming regulation of the balance between men and women."[12] It is sad enough to live in a society that denies the equality of women in many ways. It would be even sadder if a situation developed that even denied women an equal opportunity of being born.

THEOLOGICAL CONSIDERATIONS

Very few biblical principles apply to this area of artificial sex selection. Apart from the obvious concern for using abortion for sex selection, there are few clear guidelines. Most comments are intuitive and open to much discussion.

Although artificial sex selection may be beneficial in limiting populations in underdeveloped countries, there appears to be little warrant for its use. If it is implemented solely to cater to the whims and desires of parents, then Christians should oppose its use. There is already a definite trend toward deliberate parental control of the child to be conceived. Children are a gift from God (Ps. 127:3), and to tamper with their genetic structure seems dangerous. It is one thing to remove genetic defects brought on by the Fall. It is quite another to specify sex or other genetic preferences. Therefore, it seems that while artificial sex selection may be permissible as a means of limiting births in third-world countries (although even that use is questionable), it would be strongly inadvisable to promote it world-wide.

IV
In-Vitro Fertilization

When Aldous Huxley wrote his famous book *Brave New World* in 1932, very few people thought that what he predicted would ever take place in their lifetimes. Louise Brown—the world's first test-tube baby—was born on July 25, 1978, and now the technological part of the book seems within our grasp. Some have been concerned that the entire *Brave New World* scenario might take place in the future and bring great harm to society. Others are encouraged about this new advance and feel it will not upset societal patterns.

In-vitro fertilization (IVF) or ectogenesis is a procedure that allows an egg to be fertilized and grown outside the womb for a short period of time. It is not yet possible to grow a test-tube baby completely outside of the womb. Further advances in the development of an artificial placenta are necessary for this to take place.

IVF is accomplished through the following process (see Figure 1, p. 54). First, the woman is treated with a follicle-stimulating hormone to stimulate the maturation of the eggs in her ovary. Second, an incision is made in her abdomen and the eggs are removed with the help of laparoscopy. A laparoscope (a long metal tube with fiber optics illumination) is fitted with an attachment to allow the removal of ripe eggs by suction.

Third, the eggs are placed in a dish containing blood serum and nutrients. The eggs are fertilized in the dish by sperm. Before the sperm is introduced, it is diluted in order to simulate conditions in the Fallopian tubes and undergoes a process called capacitation that prepares it to fertilize the eggs. Fourth, the fertilized eggs are transferred to another dish of nutrients and allowed to divide. Transfer of the eggs takes place at the morula stage (eight-sixteen cell stage) or the blastocyst (fifty-sixty cell stage). The woman is then given further hormonal treatment to prepare the uterus and one of the blastocysts is placed in the uterus through the use of a plastic canula. In a successful attempt the embryo attaches to the uterine wall, and further development takes place in the mother.

Figure 1

IVF FERTILIZATION

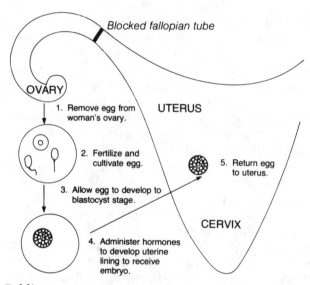

Public concern about IVF research surfaced in 1972 when the issue of fetal research was being criticized by a number of ethicists and theologians. Essays in the *New England Journal of Medicine* and the *Journal of the American Medical Association* were particularly important in focusing public and medical attention on IVF research.[1]

These articles were criticized and arguments for IVF were presented by theologian Joseph Fletcher in his book *The Ethics of Genetic Control* and biologist R. G. Edwards, who published a survey of various questions surrounding IVF techniques in the *Quarterly Review of Biology*. Edwards had previously published an article in *Nature* with David Sharpe listing the benefits of IVF and advocating interdisciplinary consultation as the best method for social monitoring of the research.[2]

Governmental involvement in IVF resulted from a larger concern over the question of fetal research, research with pregnant women, and research on children. In 1974 Congress established the National Commission for the Protection of Human Subjects, addressing a number of areas of research but not specifically dealing with IVF research. The Department of Health, Education, and Welfare (HEW) established a moratorium on IVF in 1975 with a procedural requirement that any application or proposal involving IVF must be reviewed by the Ethics Advisory Board, which was to render advice as to its acceptability.[3] Deliberations by this board were triggered when HEW received an application for support of IVF research in 1977 from Dr. Pierre Soupart of Vanderbilt University.

Before the board could meet to evaluate the request, Louise Brown's birth was announced. With the publicity of the birth of the first test-tube baby came increased attention on the subject of IVF in the United States. The board met to consider the Soupart application and held public hearings in the ten regions in order to stimulate debate. After these and other deliberations, it published its report and conclusion in 1979.[4] The board concluded that research in external fertilization and embryo transfer would be accepted. Thus the ban was lifted, but federal funds have not been forthcoming.

Debate in this country was rekindled with the birth of the first American test-tube baby—Elizabeth Jordan Carr—on December 28, 1981. With the birth of more than twenty IVF babies world-wide, greater attention has been given to this technique.

SCIENTIFIC CONSIDERATIONS

There are a number of scientific issues that affected the decision of HEW and other bodies to regulate IVF research. The primary concern was to determine whether IVF presented a greater risk than normal processes of fertilization. This was difficult to ascertain since the amount of experimentation in this field has been extremely limited.

The revised Declaration of Helsinki (1975) notes that biomedical research involving human subjects should be based on adequately performed laboratory and animal experimentation. Following the spirit of this declaration, Leon Kass, Marc Lappe, and Benjamin Brackett called for further animal research with IVF before clinical application and embryo transfer in humans was performed.[5]

Although some studies with animals have been done, they are not sufficient. From 1959 to the present there have been thirteen papers on IVF and embryo transfer in rabbits, eight papers for mice, and one for rats.[6] Even fewer studies have involved nonlitter-bearing mammals. For example, at about the same time Australia was announcing the birth of its seventh test-tube baby, Dr. Benjamin Brackett at the University of Pennsylvania was announcing the birth of the first calf conceived in a test tube.[7]

The following comment by Dr. Martin M. Quigley of the University of Texas Health Science Center (Houston) is appropriate here:

> Despite the fact that many animal studies have been done with this technique, we know very little about the abnormalities it can produce. This is because most studies are concerned with achieving fertilization and do not go on to transplant the fertilized embryo in a host mother to see what type of offspring might be created.[8]

Such prior experimentation is necessary to determine the potential risk of gross and genetic abnormalities that might result from the artificial removal, fertilization, growth, and implantation of fertilized ova. It is especially important to have prior primate experimentation that could help assess the risks and effects *before* they occur in human subjects. Dr. Luigi Mastroianni, chairman of obstetrics and gynecology at the University of Pennsylvania, said, "It is my feeling that we must be very sure we are able to produce normal young by this method in monkeys before we have the temerity to move ahead in the human."[9]

When a moratorium was called on human experimentation involving IVF in 1972, Steven Hecht and Marc Lappe wrote a letter endorsing the moratorium in the *New England Journal of Medicine.*[10] They noted that experimentation could be done at the Southwest Foundation for Research and Education where there were 150 breeding baboons. Such a study would have been extremely helpful in determining the possible effects of IVF in humans and could have been accomplished by monitoring the rate of neonatal deformity in the fetuses. However, no such experimentation was ever done.

Such experiments would not have completely ruled out the possibility of abnormalities. It would still be possible that certain species differences in sensitivity to the physical manipulations or to the nutrients used to grow the eggs could result in fetal abnormalities. These studies also would not have ruled out the possibility of mental retardation or other abnormalities that would be difficult to assess in primate subjects. Nevertheless, primate experimentation would be the very least that should be done to insure the safety of unborn children.

The fact that such experimentation did not precede the births of the first test-tube babies raises serious scientific, as well as ethical, questions. Although these first few children seem normal, this does not dismiss the concern for possible abnormalities. All we can fairly conclude is that IVF techniques do not seem to produce gross abnormalities (like those associated with drugs such as thalidomide) but we cannot entirely eliminate the possibility that there might be genetic abnormalities.

There should still be some concern as to whether IVF has more subtle effects. Many factors such as mental development or late-onset medical abnormalities cannot be predicted. In a sense the first test-tube babies will serve as test organisms. They have become the *means* by which scientists perfect their techniques rather than the *recipients* of carefully researched procedures. It is nearly impossible to guarantee that they won't later die from some genetic disorder precipitated by IVF techniques. Harvard biologist Ruth Hubbard has said:

> The guinea pigs are the children who are produced, for none of them can we have adequate animal models. They cannot consent to be produced. And we cannot know what hazards their production entails until a large enough number of them will have lived out their lives to allow for statistical analyses of their medical histories.[11]

When the risk of fetal abnormality is present, it is unwise to proceed with IVF on humans. Standard medical procedure should be to move cautiously from one experimental step to another. The possibility of defects raises a serious moral objection. Leon Kass has said:

> This moral objection should be widely shared, for it rests upon that minimal principle of medical practice, do no harm. In these prospective experiments upon the unconceived and the unborn, it is not enough not to know of any grave defects—or at least no more than there are without the procedure. The general presumption of ignorance is caution. When the subject-at-risk cannot give consent, the presumption should be abstention.[12]

The techniques used in England raise even greater concerns because of their use of hormones. Hormones are injected to prepare the womb and maintain the pregnancy. In the United States such treatments might be criticized by the Food and Drug Administration because there would be a risk of producing congenital abnormalities or of producing defects in the future offspring of the test-tube children.[13]

Another major scientific issue is the concern for human life. The present techniques used for IVF are wasteful of fetal life and often result in miscarriages that may pose a threat to the mothers. Between November 1977 and August 1978 Drs. Edwards and Steptoe made thirty-two attempts to implant embryos into the wombs of mothers. Only two implants resulted in births, both of which were premature, while two other implants resulted in spontaneous abortions.[14] Dr. E. S. E. Hafez, professor of gynecology at Wayne State University, has said, "I estimate that in Sweden, Germany, France, the United Kingdom, and Australia, 20,000 women have been treated, yet we only have three proven births.[15]

In many cases of IVF a technique known as hyperfertilization is employed by which more than one egg is harvested and fertilized. One study has shown that the technique of hyperferlilization may be associated with higher rates of chromosomal abnormality (trisomy) in humans.[16] Another study done with mice indicates that chromosomal abnormalities (triploidy) can result from too high a concentration of sperm around the ovum in-vitro.[17]

There is also some evidence that IVF may by-pass the natural screening of sperm achieved by the human reproductive system in-vivo. It appears that the female reproductive tract selectively eliminates abnormal sperm so that very few reach the site of fertilization.[18] These and perhaps other factors may account for some of the loss of fetal life.

There is also the concern that the implanted fertilized ova seem to have a high rate of miscarriage or spontaneous abortion. Any technique that increases that risk also increases the risk of problems for the mother. Some of these miscarriages occurred as late as the twentieth week; so there should be great concern over the potential harm that could result from the present IVF techniques. The technique used to produce Louise Brown was only about 1 percent effective. Newer techniques are much more effective. However, even with the higher level of effectiveness, there is still a great loss of fetal life and risk to the mother. Until the effectiveness of these techniques at least equals that of normal conception, there will continue to be sufficient scientific reasons against the use of IVF.

SOCIAL AND LEGAL CONSIDERATIONS

One major legal question is the issue of surrogate mothers. Most of this discussion has centered around the question of artificial insemination in which another woman is impregnated by the husband's sperm. With the development of IVF, there is a possibility that another woman could carry the child formed from the couple's sperm and egg. This could further complicate the entire matter.

If the cause of infertility was the mother's eggs, the couple may decide to use eggs from a donor woman. This is the reverse situation from the original surrogate-mother problem created by artificial insemination. In the standard case a child is produced from the sperm of the husband and the egg of the *surrogate* who bears the child. The wife is not genetically involved. In this new twist the child is produced from the sperm of the husband and the egg of the donor, but the child is carried by the *wife*.

Most state laws do not deal with the possibility that procreation and childbearing would be separate functions. Artificial insemination, in-vitro fertilization, and embryo transfer will significantly confuse the question of parenthood. State legislatures are not considering these issues, and the courts have been reluctantly drawn into these discussions.

The legal status of IVF clinics and operations in state medical schools and hospitals is also uncertain. The most important test case in this field has been the establishment of an IVF clinic at the East Virginia Medical College in Norfolk, Virginia. In early 1979 the college announced that the obstetrics and gynecology department would offer the IVF procedure.[19] Since the clinic is privately financed, it is not affected by the federal ban of funds. It costs each patient a lab fee of $560 and a hospital operating-room fee of $1,000. If a second attempt is needed, then the cost goes to $3,120; and by the time other expense are added (such as travel and medical consultation), the total cost averages around $4,000. Even with this high price tag, over two thousand inquiries have been made to the clinic.[20]

Many couples who have come to the Norfolk clinic were hoping that their medical insurance might cover the procedure. They were disappointed when the Virginia Blue Cross-Blue Shield voted not to reconsider its policy of prohibiting payments for preconception "experimental services."[21]

There has been a long struggle over the establishment of this first IVF clinic. However, on January 8, 1980, Virginia's health commissioner approved a "certificate of need" that authorized the medical school to open the clinic at Norfolk General Hospital.[22] Attempts to stop the opening of the clinic by placing a ban on all such work in Virginia failed when the Virginia House Committee on Health, Welfare, and Institutions defeated a bill by delegate Lawrence Pratt and rejected his pleas to study it at the next year's General Assembly.[23] Legal attempts to close the IVF clinic were abandoned by the Virginia Society for Human Life in 1981.[24]

Even greater legal questions arise with the discussion of medical liability for IVF procedures. It is not hard to imagine that this new technique will ultimately lead to an entirely new area of medical litigation.

Previously, parents were not able to sue "Mother Nature" for defects in their children. What took place in the womb has, in most cases, been outside the jurisdiction of the legal system. With IVF lawsuits seem more likely to develop over defective births. When a child produced by IVF is born with some defects, it might be assumed that the doctor performing the IVF would be liable.

Two possible suits might be brought on behalf of the *child.* First, a suit might be brought on the basis of prenatal or even preconception injuries sustained by the child. The plaintiff would have to show a causal connection between the procedures of IVF and the injuries in order to collect damages.[25] Second, a "wrongful life" suit might be brought on behalf of the child. In most cases courts have refused to recognize "wrongful life" cases.

There might also be action taken to compensate the *parents.* The first would be a "wrongful birth" suit. This would be somewhat difficult to prove unless the plaintiff could show causation between the IVF and the injuries or handicap. A second would be action for the "wrongful death" of the hoped-for child. Presently, all states allow such suits if death from prenatal injuries occurs following the live birth of a fetus conceived by natural means.[26]

By extending this line of argument for "wrongful death" to include destruction of the would-be parents' property, a court case was instituted for an American attempt at IVF. Dr. Landrum Shettles secretly performed the first stages of an IVF for Mr. and Mrs. Del Zio at Columbia Presbyterian Hospital. When Dr. Raymond Vande Wiele, chairman of obstetrics and gynecology at Columbia University Medical School, discovered the vial, he ordered it destroyed because of possible fetal abnormalities. The Del Zios brought suit against the doctor, the hospital, and the medical school for $1.5 million as recompense for the destruction of their "baby."[27] A New York jury awarded them $50,000 for emotional distress following the intentional destruction of a culture containing their gametes.[28]

A final area of litigation might result from destructive genetic materials provided for IVF that did not involve the sperm and egg of the married couple. Suppose the eggs and sperm provided by a donor produce a child with a severe genetic defect (e.g., Tay Sachs, sickle-cell anemia, diabetes, Down's syndrome, Huntington's chorea). Would the parents be able to sue the donor for injury to their child or for defective materials?

The answers to such questions await future litigation and judicial decisions. One bizarre example of where this could lead can best be found in a recent court case in West Germany. A man hired a neighbor to get his wife pregnant since he was unable to produce viable sperm. After receiving $2,500 the neighbor had sexual intercourse with the man's wife three times a week for six months. She did not get pregnant. When the man found out his neighbor was sterile, he took his neighbor to court and sued him for breach of contract.[29]

Difficult legal questions will eventually involve the state legislatures and the courts to control artificial reproduction. Christians should be concerned about this because as the government becomes more involved in artificial pregnancies, it is possible that it will most likely begin to exercise greater control over natural pregnancies. The specter of governmental control in the bedroom is not a promising one.

ETHICAL CONSIDERATIONS

There are a number of different and often conflicting ethical questions raised concerning IVF techniques. In order to summarize some of the major arguments, it is best to consider the following three areas of ethical discussion.

The first major ethical concern is *the slippery slope argument*. Moral choices often lead to unintended consequences in the future. A practice that may be moral can often very naturally lead to immoral consequences. A new practice brings particular views and values with it that shape the world view of those involved so that previously repugnant practices seem more reasonable.

Already there is some concern being expressed by those involved in using IVF in a limited way. Dr. Carl Wood, a pioneer in the field of IVF in Australia, has called for a governmental investigation of the moral issues of this research. He says, "I am facing increasing demands from very imaginative and capable people who don't see any ethical problem in developing research in the test tube field."[30] In this rapidly expanding field, he feels that certain safeguards are necessary.

In the United States the Ethics Advisory Board recommended that IVF be used only for married couples. This was an important safeguard, but there is nothing in principle that would prevent the same technique from being used outside of the marriage bond.

We have already noted the possibility of establishing surrogate mothers as an alternative form of reproduction. This might even lead to an entirely new profession—prenatal baby care. This practice might be provided initially for those who could not carry a child to term, but it could easily develop into a system in which women hired other women to carry their babies while they themselves pursued their careers.

Some doctors are already working on the process of artificial embryonation (AE). In a clinic established by Dr. Richard Seed, a couple pays a fee for a woman to be artificially inseminated by the husband. After a few days the woman returns to the clinic so that the doctor can flush out the embryo and implant it in the wife.[31]

It is not unlikely that these practices could spread far beyond surrogate mothers. It could easily lead to the mixing of genetic material. For example, suppose Mrs. Brown (the mother of the first test-tube baby) would like her second child to have blond hair and blue eyes. Since her husband does not have those characteristics, she might decide to have the doctors mix her eggs with the sperm of Mr. Jones (a blond-haired, blue-eyed friend) and implant the fertilized egg in her womb or the womb of a waiting surrogate mother.

In fact, it might even be possible that only eggs would be used. Dr. Pierre Soupart at Vanderbilt University has done experiments that fuse two eggs together. The eggs are fused together as an oocyte fusion product (OFP) and begin to divide as an embryo. Since this method of reproduction would create only female offspring and require only eggs from females, it has attracted the attention of lesbian groups around the country as a possible means of reproduction for lesbian couples.[32]

The complication continues once we add the distinct possibility that a surrogate mother might be replaced by a surrogate father. Two researchers in Melbourne predict that embryos could be fertilized in the laboratory, implanted in men's abdomens, and delivered by Cesarean section.[33] Such a procedure has already been performed by Dr. Cecil Jacobsen at George Washington University Medical School with male chimps who delivered healthy chimps via cesarean section.[34]

How far away from *Brave New World* are we? Leon Kass views IVF as "giant steps towards the full laboratory control of human reproduction."[35] We have the technology to fertilize eggs and implant them in any womb. Advances in storage techniques for genetic materials give us the possibility of mixing sperm and eggs of people from different continents or even from people who are no longer living.

Some question whether this sociological scenario need follow the technological one. In other words, just because we have the technological ability Huxley talked about in *Brave New World,* this doesn't necessarily mean our society will become like that found in his book. Can't we prevent that from taking place?

Certainly, we *can* prevent it, but there is some indication that the slide down the slippery slope is already taking place. It is not unlikely that we will soon see the development of "genetic supermarkets" in which women can enter and pick out the genetic materials they would like to have from the frozen stock and return at a later date to have the fertilized ova implanted in their wombs.

The evidence for this is convincing. First, it already has the support of respected scientists in the field. Bentley Glass, the retired president of the American Association for the Advancement of Science, remarked in his presidential address, "The way is thus clear to performing what I have called 'prenatal adoption,' for not only might the selected embryos be implanted in the uterus of the woman who supplied the oocytes, but in that of any woman at the appropriate time in her menstrual cycle."[36]

The growth of sperm banks around the country will soon be followed by the development of egg banks. Dr. Richard Seed has begun plans for the extensive use of embryo adoption (EA). Donor sperm is used, and a couple "adopts" the embryo the same way they might adopt a child.[37]

Second, such practices are already taking place with artificial insemination, thus opening the way for similar uses to be made of IVF. With the use of AID, unmarried women are having children from men they do not even know. Many lesbian couples have had children for their "marriage" by using artificial insemination.[38]

In some cases women are having children by sperm provided by men who have long since been dead. One of the more famous of these cases involved the birth of Kim Casali's third son. Mrs. Casali, the cartoonist for the "Love is" cartoon strip, had a child through artificial insemination of her husband's stored sperm[39] almost a year and a half after his death.

Third, many argue for using IVF and other artificial reproductive techniques in order to improve the genetic fitness of mankind. Eugenics has never been very popular in the United States, but it would be naïve to believe that it will not have any effect on the political arena. Leon Kass points out that the scientific reasons used to justify Dr. Soupart's research grant can also be used to justify further embryonic manipulations.[40] When these arguments are coupled with economic ones that establish the need to prevent genetic defects from increasing in our gene pool, it is likely that some policies will be implemented to produce more genetically fit individuals.

When sperm banks began to be used, an article in *The New York Times* heralded, "That day is here now, not just as a laboratory curiosity but as a commercially available sperm banking service that brings much closer the prospect of controlled breeding programs to produce superior members of the human race."[41] The recent disclosure of the sperm bank in Los Angeles for storing sperm from Nobel prizewinners would also be an example of a similar kind of hope.[42]

In conclusion, it seems likely that IVF will offer the possibilities of increasing genetic fitness, and there is some evidence that it will be used for that purpose. Already, Joseph Fletcher and others have begun to meet with state officials to urge them to establish quality control standards for reproduction, arguing that it is more benevolent and prudent both morally and economically to prevent the production of children with genetic defects.

A second major ethical concern is the *conflict between ends and means.* Most would agree that the ends of this research and the application of it are commendable. Providing children to couples who want them and removing genetic defects are worthy goals. However, moral questions are raised concerning the means for obtaining such goals.

This is one place where Christians must speak out strongly. We live in a society that thinks pragmatically and frequently neglects ethical issues. As Dr. Richard Seed of Chicago's Reproduction and Fertility Clinic says: "The public doesn't give a damn about ethics. They're dull and boring. What people care about is whether a new reproductive technology is going to do something for them or for their neighbor."[43]

Christians must raise questions about the means by

which these ends are achieved. Motivated by a technological imperative that in effect argues that if we *can* do it we *should* do it, many in the scientific and medical communities are willing to proceed without asking ethical questions. We must make sure that the sanctity of human life is not trampled underfoot in the mad rush for new reproductive technologies.

Although we may learn a great deal from experimentation on the fetus, the problem is whether this information will benefit the fetus receiving the application. In particular, there is the continuing question of consent. Prior consent is an accepted practice in most scientific and medical experimentation. The problem in this case is that consent would have to be obtained from someone who does not yet exist.

Leon Kass summarized some of these concerns:

It is one thing voluntarily to accept the risk of a dangerous procedure for yourself (or to consent on behalf of your child) if the purpose is therapeutic.... It is quite a different thing to submit a child to hazardous procedures which can in no way be therapeutic for him.... This argument against non-therapeutic experimentation on children applies with even greater force against experimentation "on" a hypothetical child (whose conception is as yet only intellectual). One cannot ethically choose for the unknown hazards he must face and simultaneously choose to give him life in which to face them.[44]

A third ethical concern, already mentioned, is *the place of reproductive technology in society*. IVF will affect future generations in a profound way. Paul Ramsey has raised the concern that "those who come after us may not be like us."[45] We may so drastically change human populations that they will have very little in common with us at all.

It is argued that such power is extremely potent and that we should not naïvely disregard its impact on society and the future of mankind. Anticipating such future advances, C. S. Lewis noted:

From this point of view, what we call Man's power over Nature turns out to be a power exercised by some men over

> other men with Nature as its instrument.... And all long-term exercises of power, especially in breeding, must mean the power of earlier generations over later ones.... There neither is nor can be any simple increase of power on Man's side. Each new power won by man is power over man as well. Each advance leaves him weaker as well as stronger. In every victory, besides the general who triumphs, he is also the prisoner who follows the triumphal car.[46]

This is not to argue against technological progress but rather for the wise use of it. The uncritical acceptance of technology by many is a major cause of concern.

It has been argued that IVF is an inappropriate technology for medical science. It does not seek to cure any disease or improve the health of the mother. Infertility is not truly a disease nor a clinical defect, and IVF is not strictly a medical procedure.

Even if one considers infertility as a medical defect, there are other means often available to provide children to couples who want them. First, there is the possibility of *tuboplasty*. This is a procedure by which the Fallopian tubes are repaired or reconstructed. These tubes may be blocked because of infections from such causes as gonorrhea and can often be repaired through microsurgery. Estimates vary as to the degree of success of this operation from 20 to 50 percent.

A second possibility is *adoption*. Current figures show that nearly three-fourths of the population is open to using adoption as a means of parenthood. One writer has estimated that there may be as many as 500,000 children eligible for adoption in this country.[47] Even if this estimate is too high, the probability of adopting a child is still much greater than having a child through IVF. While many parents would find it difficult to adopt an infant of their own race, there are many other children of older ages and different races who are available for adoption. Unfortunately, our more liberal abortion laws have made fewer children available for adoption today, and in a sense IVF has become a means by which to circumvent the problems brought about by abortion.

A third option is *drug treatment.* Many times female infertility is a result of a failure to ovulate. A drug (bromocriptine mesylate) used in Europe to treat female infertility has also become available in this country. Although it does not cure the ultimate physical problems of the women, it does effectively treat the disorders associated with ovulation failure.[48]

A fourth experimental option is a procedure known as *low tubal ovum transfer.* A laparoscope is used to remove an egg from the ovary and then implant it by means of a syringe into a Fallopian tube on the other side of the blockage. Since fertilization is done in-vivo and hormone injections are not used, many of the scientific concerns raised about IVF are no longer a consideration.

A final option is *to remain childless.* We should look on children as "a heritage from the LORD" (Ps. 127:3). Somehow our culture has made us feel that being single is wrong and being married and childless also is wrong. Often Christians have been to blame for fostering that attitude. Yet since we live in a fallen world (Gen. 3), medical abnormalities are bound to occur. We are called to exercise dominion over the creation, but that does not give sanction to the use of any form of technology. If a medical defect can be corrected with an ethically acceptable medical technology, then such procedures should be utilized. If these are not available, then it should be no disgrace in our society for a couple to be childless.

There seems to be a tendency to want to use IVF to circumvent or manipulate the will of God. In Old Testament times Abraham and Sarah asked the Lord for a child (Gen. 18); today couples go to the Norfolk clinic. We must recognize that God is sovereign even over reproduction.

Being childless does not prevent a couple from having a ministry in the lives of children. A husband and wife without the responsibility of their own children may be able to become foster parents or spiritual parents to a large number of children. In this way their influence on lives will be even greater. A couple may also be called by God to a vocation that can be done only by a childless couple (e.g., missionary work in a dangerous place or a job requiring extensive travel). A wife without children may also be able to minister more effectively to single women. Her sympathy may be much greater for a single woman, who has neither a husband nor a child.

Unfortunately, there is a movement abreast today that argues that each couple has the right to children. If they are not granted children, it is argued, this is deprivation and might lead to a breakdown of the marriage. Having children, however, is not an ultimate right of a couple. If there is a right to childbearing, it is certainly not absolute. A couple's desire for children must be balanced by the safety of the experimental method. In most cases the rights of the unborn child are no more considered than they are in abortion questions.

This new clamoring for rights has significantly changed medical practice. Now medical science is dedicated to treating "desires" and giving people their rights rather than meeting prescribed medical ills. Paul Ramsey points out this problem:

> If medical practice has an obligation to guarantee these wishes (having a child) it may have an obligation to guarantee the other (a genetically perfect baby). In my opinion, medical practice loses its way into an entirely different human activity—manufacture (which most want to satisfy desires)—if it undertakes either to produce a child without curing infertility as a condition or to produce simply the desired sort of child.[49]

IVF is an inappropriate, and actually quite peculiar, form of reproductive technology. It is a very questionable practice of medicine. Leon Kass has said, "Just as abortion for genetic defect is a peculiar innovation in medicine ... so laboratory-fertilization is peculiar treatment for oviduct obstruction."[50]

THEOLOGICAL CONSIDERATIONS

A number of theological issues are raised by IVF research and implementation. The most important involve the biblical views of human life and the family.

The first theological consideration is *the basis for the sanctity of human life.* We are seen as special in God's creation because we are made in His image (Gen. 1:27). This is the basis for recognizing the special care and protection that must be given to the unborn child.

This view comes into conflict with a number of different IVF practices. First, there is the previously mentioned loss of fetal life. Even with the more improved technique, there is still a 90 percent loss of fertilized ova. Second, there is a general practice of destroying fertilized ova if they appear abnormal. Also, in most clinics parents must agree to abort a fetus that is found to be defective. Third, there has been the practice of hyperfertilization. Many eggs are fertilized simultaneously, one is selected for implantation, and the others are thrown away.

It should also be noted that the basic research for IVF in this country would come from the federal funding of Dr. Pierre Soupart at Vanderbilt University. His request for funds, which have never been forthcoming, was for the fertilization of hundreds of human ova to examine them microscopically for abnormalities. After his study they would be discarded, and there are no plans to even implant them in women.

Richard McCormick of the Kennedy Institute of Bioethics at Georgetown University noted this lack of concern for the fetus:

> Are these really mini-abortions? The evaluation of nascent life in these early days is indeed a problem. But that does not mean the problem can be decreed out of existence by simply going ahead. Where human life is at stake and we have doubts about its evaluation, does not prudence dictate that as a general rule life enjoy the benefits of our doubts?[51]

If we are uncertain of the moral status of the early developing embryo, then it would seem wise to provide protection for it. Ethicist Allen Verhey has pointed out that

> even if one did not hold that the human being's history begins with conception, respect for life is nevertheless violated here. Respect for life is violated because here (unlike the situation of abortion) human life is created in order to be destroyed. Here the procedure demands from the very beginning the intention to kill those intentionally fertilized but not chosen.[52]

Although some practices that are inherently wasteful (e.g., hyperfertilization) can be curtailed, much of the concern will still remain because IVF techniques are not perfected to the point where safety can be guaranteed to the fetus.

A second theological consideration is *the effect IVF will have on a biblical view of marriage.* IVF separates the physical dimensions of sexual intercourse from the emotional and spiritual ones. Procreation becomes reproduction (in the mechanical sense) and turns the marriage bed into a chemistry set.

Human parenthood involves two spheres: the unitive (Gen. 2:24) and procreative (Gen. 1:28). These are tied together by the union of sexuality, love, and procreation. Making love and making babies (to use the vernacular terms) are tied to the same physical act. The pleasure of sex, the communication of love, and the desire for children are unified in the same act. IVF separates these functions and thus poses a potential threat to the completeness God intended for marriage.

Motherhood will also be affected. Childbearing would no longer be a natural outcome of procreation. The proliferation of surrogate mothers will continue to blur the true relationship between procreation and parenthood and could therefore lead to a breakdown of the family unit. God intended that the family thrive (Eph. 6:1–4; Col. 3:18–21), and current genetic advances certainly pose a threat to the stability of the family.

The stability of a family is not dependent on having children. While God's plan is for us to reproduce our lives in our children, we can still experience God's grace without children. We live in a fallen world that occasionally prevents us from having children. Our responsibility is to use technology safely in order to redeem the effects of the Fall while acknowledging that God determines birth (Gen. 4:1; 17:16; Ruth 4:13) and is still in control, even over barren wombs (Deut. 7:14). Childless women are not displeasing to God, as is attested by the testimonies of Sarah (Gen. 18), Rachel (Gen. 29–30), Hannah (1 Sam. 1), and Anna (Luke 2:36–38). Even if our technology fails, God is still in control and can bring great blessing out of the heartbreak of infertility.

A third theological concern centers around *a biblical view of technology.* Should man be involved in genetic manipulations that could affect future generations adversely? While it is true that we have a cultural mandate to "subdue" the earth (Gen. 1:28), this must be tempered with a cautious concern about the implementation of technology.

Using technology does not mean that we are granted the privilege of playing God. Our advance in the knowledge of reproductive biology also means that we now have a greater responsibility. Reproductive biology seems to be an area that we should not try to control without greater wisdom and safeguards. Paul Ramsey makes this point by comparing this concern to the environmental movement. He points out an interesting irony:

> So today we have the oddity that men are preparing to play God over the human species while many among us are denying themselves that role over other species in nature. There is a renewed sense of the sacredness of groves, of the fact that air and streams should not be violated. At the same time, there is no abatement of acceptance of the view that human parenthood can be taken apart and reassembled.[53]

If we are to be stewards of the creation (Gen. 1:28), then we must exercise greater care than we have shown in environmental areas. If our track record has been one of raping the resources, then why should we expect that greater care and compassion will be shown to the unborn child?

In summary, we must conclude that although IVF provides a means by which to help infertile couples, it is fraught with many scientific, social, ethical, and theological problems. Further research may solve many of the scientific concerns; but unless there is a radical change in the attitudes of our world, the social and ethical ones will remain. There are options other than IVF for infertile couples; therefore IVF seems an inappropriate technology for medical science at this time.

Genetic
Manipulation

V
Recombinant DNA Research

Recombinant DNA research may change the whole land-scape of genetic engineering. Biblical critique of this technique however, has been very sparse. Christians have often been too quick to condemn genetic research and have failed to provide meaningful critique of these new developments. While there is some reason for concern, recombinant DNA (rDNA) research promises many benefits to our world.

Recombinant DNA research began in the early 1970s when these new genetic techniques were developed by biochemists in California working independently at Stanford University and the University of California at San Francisco. By using restriction enzymes, they cut small segments of DNA called plasmids and inserted foreign DNA into *Escherichia coli (E. coli)* and thus created an entirely different creature (see Figure 2, p. 78). These new creatures are actually genetic hybrids of two different organisms and have been called DNA chimeras by their inventor Stanley Cohen because they are conceptually similar to the mythological Chimera (a creature with the head of a lion, the body of a goat, and the tail of a serpent).

This technique for genetic manipulation is fundamentally different from other forms of genetic breeding that have been used in the past. Breeding programs work on existing arrays of genetic variability in a species, isolating specific genetic traits through selective breeding. Recombinant DNA techniques, on the other hand, allow scientists to, in effect, "stack" the deck of traits or produce an entirely new deck. Organisms can be redesigned genetically and new forms of life can be created in the laboratory.

Figure 2

rDNA RESEARCH

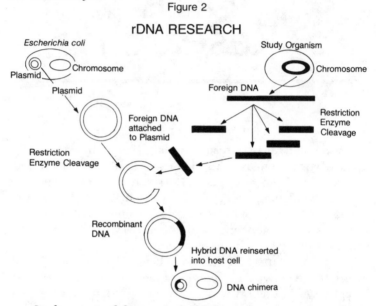

Such a powerful genetic tool raises many questions about its use. Some people are concerned that certain genetic "sleights-of-hand" might produce dangerous consequences. Ethan Singer has said, "Those who are powerful in society will do the shuffling; their genes will get shuffled in one direction, while the genes of the rest of us will get shuffled in another."[1]

There has also been the concern that some well-meaning scientist may accidentally create an Andromeda strain not unlike the one envisioned by Michael Crichton in his book by the same title.[2] A particular microorganism might inadvertently be given the genetic structure for some pathogen for which there is no known antidote or vaccine. Various forms of life might be created that would reduce the productivity of existing plant or animal food sources.

A HISTORY OF rDNA RESEARCH

When rDNA techniques were developed, there was an almost immediate concern about its safety. In the summer of 1972 Robert Pollack was teaching a laboratory safety seminar for cancer researchers at Cold Spring Harbor Laboratory. During the seminar he learned of experiments proposed at Stanford to combine DNA from simian virus 40 (SV-40) into the bacterium *E. coli*. This relatively simple virus, found in monkeys, is used in cancer research. Although it does not appear to produce cancer in humans, it can turn in-vitro laboratory preparations cancerous and thus there is still some concern about its use. Pollack called Paul Berg at Stanford to express his concern that if the altered *E. coli* escaped, it could end up in the intestine of every living person.

Although the Berg experiment was canceled, scientists were concerned that such experiments might proliferate. Herbert Boyer at the University of California at San Francisco found an enzyme that cut DNA so that it left overlapping "sticky" ends. This made rDNA research much easier. Stanley Cohen at Stanford University also discovered a plasmid named pSC (Cohen's initials) 101 that could easily slip new genes into *E. coli*. As scientists began to request these materials for their research, it became apparent that the safety issue had to be resolved or else research even more dangerous than Paul Berg's might be implemented.

A moratorium on this research was called until scientists could assemble at the Asilomar Conference on Recombinant DNA Molecules (February 1975). They drafted a report that established four levels of risk and four types of experiments.[3]

These recommendations in effect postponed most of the proposed experiments. Even moderate-risk physical containment was more stringent than standard laboratory procedures (air leaving the laboratory had to be blown through an expensive filtering system, and experiments had to be done in special hoods with a constant curtain of air). The recommendations also called for biological containment by using special forms of *E. coli* that could not inhabit the human intestine and could not even live outside the laboratory for an extended period of time.

Even with these stringent guidelines, growing concern over this new genetic technique began to mount. Politicians, scientists, and governmental agencies began to call for greater control. They were uncomfortable with scientists making policy and regulating themselves. As Erwin Chargaff of Columbia University later sarcastically put it, "This was probably the first time in history that the incendiaries formed their own fire brigade."[4]

Government involvement, however, was also criticized. The day after the Asilomar conference, the first meeting of the National Institute of Health's Advisory Committee was held. David Hogness, a Stanford scientist, was chosen as chairman of the subcommittee to draft rDNA guidelines.[5]

His selection was criticized since he was involved in very controversial "shotgun" experiments in which pieces of DNA were randomly cut up and placed in various recombinants. There was great fear that a cancer gene might accidentally be implanted in a bacterial cell that could then escape from the laboratory. One commentator felt that having Hogness write the guidelines was like "having the chairman of General Motors write the specifications for safety belts."[6]

After much criticism and revision, the NIH committee developed final guidelines.[7] They developed guidelines for physical containment (P1, P2, P3, and P4 facilities) and biological containment (EK1, EK2, and EK3 types of experiments) that were similar to those developed at Asilomar. Roy Curtiss at the Alabama Medical Center was able to develop a weakened strain of E. coli that met the biological containment criteria and was specified for certain types of experiments.

At the same time Cambridge, Massachusetts, Mayor Alfred Vellucci learned of Harvard University's plans to construct a physical containment facility for rDNA research. He, the City Council, and some of the Harvard faculty were concerned that there might be possible hazards to the community from the facility; and so two public hearings were called. At the first public hearing, Mayor Vellucci expressed his concern that "something could crawl out of the laboratory, such as a Frankenstein"[8]; and he called for a two-year ban on all research.

At the second meeting the City Council voted down Vellucci's resolution. Instead, they put a "good faith moratorium" into effect and established the Cambridge Experimentation Review Board (CERB). It was composed of eight citizens who used a jury trial format to investigate the potential hazards of the research and later issued more stringent guidelines.

Although congressional action was taken on this issue, legislation has never been enacted. Bills by Senator Edward Kennedy (S 1217) and Congressman Paul Rogers (HR 7897) died in committee, and NIH has been involved in most regulation.

In recent years the NIH guidelines have been relaxed significantly. Experimental evidence has shown that many built-in genetic features prevent the Andromeda strain scenario from being realized. In fact, it was found that many of these recombinations are already occurring in nature and do not seem to pose any danger.[9]

Revisions of the NIH guidelines call for exemptions on about 80 percent of all experiments.[10] There had also been a move to make the regulations merely voluntary. Paul Berg said he felt that such research appears to be so safe that federal guidelines should be dropped if no new problems arise.

There has been some general support for relaxing these guidelines. But some concern has surfaced recently because of published reports that a scientist accidentally used a rare insect virus prohibited by the NIH guidelines.[11] Most scientists and bureaucrats, however, feel that rDNA research does not pose a grave threat to life. The initial fears have been shown to be largely unfounded and greater freedom has been given to scientists working in this field.

POTENTIAL BENEFITS OF rDNA RESEARCH

The promise of rDNA research and its application are very great. Many Christian writers have been guilty of over-reaction against genetic engineering of any kind and have overlooked the potential benefits of this research. If used wisely, this technology can aid us in many ways.

First, it offers the potential of redesigning microorganisms so they will produce *medically important substances.* One of these is insulin. There are over a million diabetics in the United States who need to take daily doses of insulin.[12] Insulin is presently extracted from the pancreas of pigs and cows, which makes it both expensive and often dangerous to the 5 percent of all diabetics who are allergic to animal proteins.[13]

Recombinant DNA research has allowed scientists to implant a human gene for insulin into *E. coli* so that it now produces human insulin along with its other biochemical substances. Eli Lilly and Genetech, Inc., have developed this biosynthetic insulin and volunteer patients around the country are receiving injections as part of a nationwide test. Up to three hundred diabetics will begin these treatments at six medical centers including Henry Ford Hospital in Detroit and Mercy Hospital in Trenton, New Jersey. If the tests run smoothly, approval by the Food and Drug Administration should come by 1984.[14] Eli Lilly is building two plants, one in Indianapolis and one in England, in order to increase the volume and further reduce the cost of manufacturing.

Not only will such study be of great benefit in the area of applied research but also in basic research. Splicing genes together in this way should help scientists to understand more fully the complex chemistry of diabetes. It is not just one disease but a variety of metabolic disorders that affect as many as ten million people in the United States.

Biosynthetic human growth hormone has been developed by using rDNA technology. Stanford University pediatricians have begun tests of its effectiveness on a group of twenty children suffering from hypopituitary dwarfism (they are currently testing the hormone for safety in healthy adults).[15]

An estimated half a million children suffer from this growth defect that can be corrected only with this hormone. Previously, the only source for this hormone was cadavers. A year's treatment for one person required fifty cadavers. Development of the biosynthetic substitute not only brings the promise of treating dwarfism but also of treating many other disorders. It is a basic hormone that regulates a number of body functions and can be used in the treatment of obesity, severe burns, bleeding ulcers, and broken bones.

The potential of rDNA research in the field of immunology is also very great. Antibiotics are used to treat bacterial illness and can do this very safely. But in order to protect organisms from viral diseases, doctors must inject a killed or attenuated virus into the patient. This not only stimulates the production of antibodies against the virus but also carries the possibility of producing the disease in the patient. By using rDNA techniques to take apart a toxin gene, scientists will be able to develop a viral substance that triggers the production of antibodies without having to fear the possibility that the injected patient would contract the disease. Using this technique, British scientists have already created a new flu vaccine by inserting the influenza genes into bacteria, thus paving the way for future "safe" vaccines against any disease caused by influenza.[16]

Scientists in Switzerland have been successful in planting interferon genes into *E. coli* so that it will produce this natural virus fighter. Although it has not yet proven to be effective, interferon may help treat flu, hepatitis, and multiple sclerosis. Until now the use of interferon has been highly impractical since it can be extracted only in small amounts from such sources as white blood cells and thus is extremely expensive (approximately $20,000 per patient).[17]

In the United States biosynthetic interferon was developed by Genentech, Inc., and produced by Hoffman-LaRoche, Inc. Tests of the substance for approval by the Food and Drug Administration are being done at the M. D. Anderson Hospital and Tumor Institute in Houston and at Stanford University.[18] Other companies are also planning tests of their own biosynthetic interferon in order to determine its safety. Although none of the tests are designed to assess the effectiveness of interferon as a cancer-fighting agent, there are other tests being conducted by the various cancer agencies to test its effectiveness.

Other medically important substances that could be produced are human growth hormone, clotting factor VIII (for hemophiliacs), fibrolysin, urokinase, lysosomal enzymes, and various antibodies and antigens. Bernard Davis has even suggested that rDNA research could help in the production of specific cancer antigens.[19]

A second benefit of rDNA research will likely be the *improvement of plant species*. This technique could help develop more efficient biochemical pathways of photosynthesis and thus increase crop yields. There is also great promise in the use of rDNA techniques to develop plant strains that can resist viruses and are less susceptible to herbicides. It could also help reduce the present demands for nitrogen fertilizer.

Certain plant species have the property of nitrogen fixation. They are able to take nitrogen molecules in the air and convert them to ammonia, which is then used in plant growth. Most crops do not possess this ability, and nitrogen needs to be provided artificially through the application of synthetic ammonia fertilizers. This process is not only time consuming but very expensive and pollutes surrounding water sources.

By using rDNA techniques it should be possible to alter the genetic structure of the plants themselves so they can convert nitrogen to ammonia with their own biochemical machinery. Already, researchers at Cornell University have transferred the genes for nitrogen fixation to yeast cells. As the genetic structure of commercial crops is known, we may be able to apply this technology. One problem is that this nitrogen fixation capability might reduce crop yields because of the additional energy needed for this capability.

A better method of providing nitrogen for these crops might be to redesign the bacteria responsible for nitrogen fixation. In the legumes there are bacteria that live in a symbiotic relationship with the plants in nodules in their roots. Genes necessary for nitrogen fixation might be placed into bacteria that already live in a symbiotic relationship with existing food crops. This would reduce the need for synthetic fertilizers. The *nif* genes (nitrogen fixation genes) have already been isolated; therefore it is possible that this modification could be done in the near future.[20]

Because the process of nitrogen fixation is a relatively complex process (governed by at least seventeen genes), most researchers feel that the first practical benefits in agriculture will be in other areas. For example, at the University of California at Davis scientists have endowed a tomato plant with the ability to grow in sea water. At another lab in Illinois, scientists have produced a strain of corn that can survive for three weeks without water. Using cell fusion, a cross between a potato and a tomato has been made that will endow a potato with the tomato's resistance to late-season plant blights.[21]

Such advances will help farmers throughout the world. The Green Revolution agriculture exported to many other countries was too dependent on fossil fuels and petrochemical fertilizers to be of much help. Advances such as these achieved by genetic manipulation will greatly extend the area of agriculture without the previous Green Revolution requirement of fossil fuels.

A third benefit of rDNA research will be to *improve industrial processes* that presently use microorganisms. Genex, a Bethesda, Maryland, firm active in this field, has identified over one hundred products that will probably be manufactured more easily by using rDNA technology.[22] Industries that manufacture drugs, plastics, industrial chemicals, vitamins, and cheese are just a few examples of industries that will benefit from this technology.

Monsanto has shown interest in using this technology to produce ammonia. This company believes that the use of rDNA technology will provide greater energy savings. The current method for the manufacture of ammonia requires thousands of pounds of pressure at temperatures over 500 degrees Fahrenheit. A biological method of manufacture would mean that bacteria could do the same thing at room temperature and pressure.[23]

A fourth benefit would be to design microorganisms to *perform environmental tasks*. Some could be used in energy production. Algae could be produced that efficiently split water into hydrogen and oxygen for use as fuel. Altered bacteria could more efficiently digest sugar in order to produce alcohol that could be used in gasohol. Other bacteria could be used in the neutralization of pollutants. Microorganisms have already been created that can clean up an oil spill and can decompose the herbicide agent orange. There is great hope that these creatures could be given the ability to digest chlorinated hydrocarbons and other toxic wastes.[24]

A final benefit that should be mentioned is the area of *basic medical research and treatment*. Recombinant-DNA techniques allow scientists to isolate a specific segment of DNA and place it into bacteria like *E. coli* so that they will manufacture large quantities of a substance. This would give scientists another tool that can be used in mapping genes on chromosomes. Once this is accomplished, the possibilities of gene therapy for genetic diseases are very great.

A new advance in this direction has been announced by researchers at Yale University who injected foreign genes into mouse eggs and induced some of the embryos to incorporate these new genes into their tissues. Using standard rDNA techniques, researchers have copied genes from two viruses and injected them into mouse eggs. More recent experiments involved the successful implantation of a human gene into mice. The gene was passed on through natural breeding to two generations.[25]

Gene transplant research has been quite controversial. Experimenters at the University of California at Los Angeles announced that they had been successful in isolating a gene in mice that conferred some resistance to an anticancer drug. They extracted and copied these genes using rDNA techniques and injected them into mice that picked up the genes in their bone marrow.[26] The hope was that this technique could eventually be perfected for use in cancer therapy so that chemotherapy drugs would not destroy healthy bone-marrow cells.

Controversy broke out, however, when it was announced that experiments had been done not only with mice but also with humans. In both Israel and Italy, Martin Cline had applied this technology to patients suffering from a genetic blood disorder called thalassemia. He removed small amounts of bone marrow from the patients and mixed it with genes, that had been copied with rDNA techniques and were capable of directing the production of normal hemoglobin. Cline had previously submitted a proposal for such experiments in this country but had been turned down for lack of animal study.[27] The National Institute of Health charged that Cline violated their guidelines and asked its committees on funding to decide whether to revoke federal funding for all of his research projects.

Richard Axel of Columbia University was especially critical of the experiments. As a pioneer in the field, he pointed out that transferred genes often fail to express themselves or may run wild (producing a surplus of products) because they are not subject to normal gene regulation. He also expressed concern that these techniques were still very primitive and that a donor organism would also take up hundreds of other genes, some of which would be harmful.[28] The controversy prompted a debate concerning the ethics of gene therapy in human beings in the *New England Journal of Medicine* between French Anderson and John Fletcher of NIH and Karen Mercola and Martin Cline of UCLA.[29]

SCIENTIFIC CONSIDERATIONS

Some of the most important aspects of the rDNA debate are scientific ones. The greatest concern with rDNA research has been over its potential hazards. Once research in the field was able to lessen the possibility of potential hazards, many concerns diminished.

Scientific considerations focused around four major areas of discussion. First, there was a concern that rDNA research *would bridge the genetic barrier between prokaryotes and eukaryotes.* Prokaryotes are primitive cells (like bacteria and blue-green algae) that lack a nuclear membrane, while eukaryotes are fully nucleated higher organisms that have a much different and sophisticated organization.

Robert Sinsheimer noted that these organisms do not

interact with each other at the genetic level. They also have different control elements to read their genetic codes. Sinsheimer argued that putting a piece of eukaryotic DNA into a prokaryote might endow the prokaryote with eukaryotic control signals and this would be "a sort of betrayal of state secrets at the molecular level."[30]

Such concern by Sinsheimer was legitimate and was a sound basis for the initial controls on rDNA research. Further investigation has shown that hybrids between eukaryotes and prokaryotes most likely do form in such places as the human intestine but probably die out quickly.[31] Also, it was found that the difference in control mechanisms provided a natural protection against possible injection of eukaryotic mechanisms into prokaryotes.

A second concern was over *the use of shotgun experiments.* DNA of an organism was chopped into segments containing only a few genes and placed in separate bacteria that mass-produced them for further study. The concern was that since only a few genes in any organism have been mapped, most experiments involved placing unknown genes into bacterial cells.

It was suggested that one of these fragments might alter the structure of the host bacteria so that it would develop a greater potential for disease. One of the fragments might even contain genes for "a hitherto repressed tumor virus."[32] Many types of animal cell DNA contain genetic sequences that are the same as those found in RNA tumor viruses; therefore these concerns were legitimate.

The possibility of disease resistance was also considered. The indiscriminate use of antibiotics on a world-wide scale already threatens the usefulness of many drugs. The recombination of genetic segments that confer antibiotic resistance may unintentionally be passed from one organism to another. Since antibiotic markers are often used in research, these might be passed to the host organism.

Further research, however, has shown that these concerns were exaggerated. It has been calculated that the probability of isolating a complete fragment of a gene was exceedingly small, and mechanisms were found that prevented the expression of them. Eukaryotes consist of long runs of genes with intervening sequences that govern protein synthesis (at the mRNA stage). Bacterial cells that have a single DNA strand without these intervening sections will not express the genes because pro-karyotes do not utilize these intervening sequences.[33] Thus it seems that the barrier between prokaryotes and eukaryotes actually acts as a safeguard in these experiments.

A third concern was *the possibility of infection or the escape of altered microorganisms.* The containment facilities may have been adequate, but there was concern about human error. Many of the investigators were molecular biologists who were not used to working with dangerous microorganisms. Many of these experiments were to be done in university facilities, and so there was some concern that there would be a greater possibility of infection of the student population when many of the researchers would leave the laboratory to teach classes.

Special containment facilities were part of the answer but not all of it. Over the past thirty years there have been up to five thousand laboratory-acquired infections; and a third of these took place in laboratories with special containment facilities.[34] The major difference is that rDNA research was found to be much less dangerous than viral or bacterial research with pathogens.

Biological containment was also an important precaution. Roy Curtiss of the Alabama Medical Center became interested in working with recombinant DNA in order to study and alter *Streptococcus mutan.* Because of its ability to lodge on tooth enamel and convert sugar into an acid that leaches out miner-als, it leads to tooth decay. Curtiss was about to place this microorganism into *E. coli* but became concerned that he might inadvertently create a new type of bacterium that could colonize in the mouth and promote tooth decay at an even greater rate.

At the Asilomar conference he suggested that it might be possible to create a weakened mutant strain of *E. coli* that would not be able to survive outside the laboratory. After two years of work he developed X-1776 (in honor of the bicentennial that year), which needed additional laboratory chemicals to grow and was sensitive to ultraviolet light, bile salts, antibiotics, drugs, and household detergents.[35] These changes provided further insurance against the possibility that an escaped organism would survive and reproduce outside of the laboratory.

A fourth concern is the *response of these redesigned organisms to a different environment.* There was a general assumption that there would be an isomorphism between gene structure and gene product. In other words, if gene segment G(x) produces protein P(x) in environment E(x), then it is assumed that if G(x) is placed in environment E(y), it will either produce P(x) or nothing at all.[36]

While there is some possibility that a gene placed in a foreign environment would produce an unwanted product, the possibility of survival of a redesigned organism would be very low. These redesigned organisms would be little more than what Bernard David calls "jerry-built hybrids" and would be unable to compete with existing strains of *E. coli.*[37]

There is still some concern about the question of creating new forms of life for particular environments. The creation of entirely new forms of life is one question that has not been addressed adequately. Until we know more about the structure of niches and their place in dynamic ecosystems, a cautious approach seems warranted.

Another area not discussed by scientists concerns the assumptions used to draft the guidelines for rDNA research. Both the Asilomar and NIH guidelines constructed a safety scale based on evolutionary theory. In other words, experiments performed with DNA from animals considered closer evolutionarily to man were believed to be more dangerous than those performed with DNA from less-advanced animals.

Since we live in a created world, such a rating may not be completely accurate. Certainly, similarity in morphology implies some similarity in genetic structure. We might expect that DNA sequences of other mammals would be more similar to human DNA than to viral RNA, but there most likely are some exceptions to that general rule.

One of the great dangers of the guidelines was the inherent "psychology of containment." The range of containment facilities from P1 to P4 contributed to this. A P2 facility seemed safer than a P1 because it was a step higher on the scale even though it was just a normal laboratory environment. Further problems resulted from the built-in deception of the guidelines. A scientist might be misled by the distance of genetic relationship between humans and a "primitive organism" to use a P2 facility when in actual practice a P4 facility may be needed. Scientists will need to be guided by the accurate genetic mapping of organisms and not base their decisions on evolutionary assumptions.

One might ask, If evolution is a poor standard for the guidelines, what should be put in its place? At the present time there is nothing else that can be used. Scientific research in the area of genetics continues to proceed on evolutionary assumptions. Research that creationists might propose to provide better criteria has neither been funded nor considered. Evolution is assumed to have occurred, and so alternative criteria for judging the safety of genetic research await future interest.

In conclusion, we should note that grave scientific concerns were raised during the initial stages of rDNA research. Scientists showed responsible behavior by attempting to formulate guidelines for research while the hazards were still unknown. Unfortunately, they relied too heavily on evolutionary assumptions that kept the guidelines from being what they should have been.

LEGAL CONSIDERATIONS

Some very crucial legal questions have surfaced from the application of rDNA research. The major ones have centered around the question of whether the technique and its applications could be patented.

One patent application filed by Stanford patent officer Niels Reimeyers for Stanley Cohen and Herbert Boyer was for the patent rights to the rDNA technique they invented. Another patent application by Upjohn was filed in 1974 for a strain of bacteria that produced the antibiotic lincomycin.

A third patent application was filed by General Electric for an oil-slick eating *Pseudomonas* developed by Ananda Chakrabarty. The new strain developed by him can completely digest crude oil. Although no naturally occurring organism can eat oil, there are a number that digest different parts of hydrocarbons (aromatic, aliphatic, terpene, polynuclear aromatic, and cycloparaffin). Using a technique similar to rDNA, Chakrabarty was able to join genetic material from different bacteria that could digest different parts of oil in the creation of this "super bug.[38] Scientists hope to use these bacteria to clean up oil spills. As oil begins to leak into the environment, a helicopter could come by and drop a load of bacteria into the spill.

In addition to the question about whether this creature should be patented, there is the question of safety. Is it possible that these bacteria might digest other oil? No one would welcome the spread of an infectious disease that destroys car, truck, and airplane lubrication systems. Neither would we welcome the prospect of having to pasteurize all of our petroleum products.

The recent Supreme Court decision that allowed General Electric to patent this organism is extremely important for many of the industries that use microorganisms.[39] This decision will undoubtedly lead to the vast production of novel life forms. Irving Johnson, vice president for research at Eli Lilly, commented on the impact of rDNA techniques: "You're talking about a process that could affect all living species. I can't think of a single event that is as broad as that, except maybe the discovery of atomic particles."[40]

The legal path to the Supreme Court decision was quite rocky. The Upjohn and General Electric patent applications were originally denied by the patent office's board of appeals. The reason given was that patent statutes did not allow bacteria to be patented.[41]

Both decisions were independently appealed to the Court of Customs and the ruling was in favor of the claims. The patent office, however, appealed the case to the Supreme

Court. After some postponement, the Supreme Court finally ruled on the case.

Writing for the majority, Chief Justice Warren Burger said that a living organism was covered by the federal patent law's definition of "manufacture" of inventions or discoveries. Many commentators were concerned that such a decision would virtually open the door to patent applications for any life form, since a life form was simply redefined as a combination of physical and chemical arrangements. Justice Burger argued, however, that the issue was "not between living and inanimate things, but between products of nature—whether living or not—and human-made inventions."[42]

The Supreme Court decision also affected the other one hundred patent applications awaiting the General Electric appeal. Six of the applications were filed by Eli Lilly Company and Genentech for their human insulin bacteria. Another was an application filed by the National Distillers Corporation and Cetus for a microorganism that ferments corn starch or corn syrup into alcohol.[43]

The decision will likely increase the amount of genetic research by industries. Thomas Kiley, a vice-president of Genentech, said, "We believe that court's landmark decision will spur the genetic industry. It is particularly important to smaller companies, like Genentech, who can be protected by patents so that they can enter fields such as pharmaceuticals which have been dominated by large firms."[44] Ronald Cape of the Cetus Corporation, the leader in genetic research, has argued that patents are necessary in order to protect corporate secrets and in order to provide a profit motive for genetic application.[45]

While many in the business field supported the decision, there are a number of scientists who have expressed concern over the ruling. Many scientists felt that it would prostitute scientific research in the field of genetics. The profit motive, for example, has already changed the manner in which scientific breakthroughs have been announced. Spyros Andreopoulos of the Stanford University Medical Center News Bureau lamented the recent trend of "Gene Cloning by Press

Conference" in which announcements of genetic advances were made at news conferences instead of in scientific journals in order to boost company stock and gain additional investments in various research projects.[46] The Security and Exchange Commission had to temporarily hold up the offering of Genentech stock for public sale because of the inordinate amount of publicity surrounding it.[47] With some investors saying that it could be the next Polaroid or Xerox, it is not surprising that stock was selling at as much as eighty-nine dollars a share.[48]

It was also felt that the decision by the Supreme Court would result in a decrease in the relatively free flow of scientific information. Scientists-turned-entrepreneurs sometimes have refused to share their findings for fear of commercial loss. Jonathan King of the Massachusetts Institute of Technology warned, "Now you have the prospect of keeping a strain [of bacteria] out of circulation until you have the patents."[49]

There may also be some concern that the whole process of genetic research may be hindered by the movement of universities into the arena. The process of rDNA manipulation was patented by Stanley Cohen and Herbert Boyer, who turned over royalty rights to their respective universities.[50] Although both universities have stated that they will charge minimal fees for patent use, what is to guarantee that Stanford or UCSF will not increase the royalty payments in the future in order to fund various university projects? There is already ominous news that they are seeking a second patent to acquire licensing rights over all products produced by technology covered by the first patent.[51] It has recently been reported that the University of California is threatening legal action against two drug firms for "unauthorized use" of a scientific discovery. More litigation is almost sure to follow.

Harvard University has announced its decision to transfer rDNA technology to the marketplace. It is presently considering becoming a shareholder in a company working in the rDNA area by using funds from the Harvard endowment.[52] It is ironic that the university that initially created the greatest furor over rDNA research is now one of the first to push for its transfer into technologically lucrative applications. Will this have some effect on safety standards that now might be relaxed in order to facilitate quick deployment of genetic breakthroughs into the marketplace and faster returns on investment?

Another major concern over the Supreme Court decision is that it may create numerous legal problems. In particular, there will most likely be attempts to get around patents by making slight variations of a particular organism. James Watson noted, "It will be awfully hard to show uniqueness, to prove one man's microbe is really different from another's."[53] There is some concern not only over the legal hassles that might ensue but also over the likely potential for wasted research. Many scientists might be involved in working to modify existing bacterial strains in order to break patents and so be wasting both time and research money.[54]

ETHICAL AND THEOLOGICAL CONSIDERATIONS

The major issue is the question of scientific responsibility. When the safety of any form of technology is in question, it is extremely important that scientists seriously consider the safety question and make public their concerns before research is begun and policy is formulated.

In most cases scientists acted responsibly in their discussions of the rDNA research. Although some scientists felt governmental intervention was unnecessary, the potential hazards were great enough to warrant it. Freedom of scientific inquiry should be allowed, but that does not give license for any form of scientific experimentation.

Scientists are often willing to take on such potential risks, arguing that they are the ones who will be exposed to them and they have voluntarily consented to conduct the experiments. There are examples of this in research fields as diverse as high-energy physics and biomedical research. The crucial distinction, however, lies in the certainty of not propagating the risk to others. Endangering the lives of others who have not consented to engaging in such a risk violates established ethical principles. Erwin Chargaff has said, "I may damage myself as much as I want but not one iota of danger to others is permissible."[55]

The danger in these experiments is the tendency for scientists to feel that they alone are taking on the risk. This psychology of safety is extremely dangerous. It is not only the experimenter who is taking on risks; it is also society at large. If a redesigned microorganism escapes, everyone will have to deal with the consequences. The number of reported laboratory-acquired infections indicates that it is important for laboratory safety committees to continue to monitor experiments. It is apparent that controlled laboratory conditions are not only for the safety of the workers but also for society as a whole.

A parallel issue of scientific responsibility focuses on the ethics of redesigning life. Should scientists be allowed to restructure life? The reflex reaction from many Christians is that scientists should never tinker with life. But this attitude fails to take into account the cultural mandate to use technology responsibly (Gen. 1:28).

Scientists already are involved in the subtle manipulation of genetic material through breeding programs. This new genetic technique, however, provides scientists with even greater powers. It has endowed them with the ability to combine genetic material from any creature with another. This is a fundamentally different level of genetic manipulation.

Ethan Singer, Massachusetts Institute of Technology biologist, has warned that initial genetic tinkering will move us toward full-scale genetic manipulation in areas outside of our expertise. First, we will create a new organism, "verify a few predictions, and then gradually forget that knowing something isn't the same as knowing everything. We will slowly move from high level containment to low level containment to large scale production to buying the hybrids at the local drug store. At each stage we will get a little cockier, a little surer we know all the possibilities."[56]

At this point we begin to see the arrogance of modern science. Out of fallen man and his evolutionary world view there develops an arrogant attitude toward life. Many scientists assume that life on this planet is the result of millions of years of chance. They therefore argue that intelligent scientists can do a much better job in sophisticated laboratories than "nature" can do alone. Recombinant DNA research gives these scientists a tool they have always desired. Julian Huxley looked forward to the day in which scientists could fill the "position of business manager for the cosmic process of evolution."[57]

Many scientists have expressed concern about the possibility of giving their colleagues such great power. In his lecture before the Genetics Society of America, Robert Sinsheimer noted that no one had given much attention to "the potential broader social or ethical implications of initiating this line of research—of its role, as a possible prelude to longer-range, broader-scale genetic engineering of the fauna and flora of the planet, including, ultimately man."[58]

Scientists are basically saying that we now have the technological ability to be gods, but we lack the wisdom and moral capacity to act like them. Given finite wisdom and sinful human behavior, the slide down the slippery slope is inevitable unless future guidelines are formulated concerning the implementation of rDNA technology. At the present time there is alarmingly little concern about this very dangerous application.

At this point it is important that Christians distinguish between two different types of research considered by scientists. The first is what we may refer to as *genetic repair*. This research attempts to remove genetic defects and develop techniques that will provide treatments for existing diseases. Applications would include various forms of genetic therapy and genetic surgery as well as slight modifications of existing microorganisms to produce beneficial products (e.g., human insulin *E. coli*). Christians can easily endorse this research as a beneficent application of technology that can remove evil and suffering brought to this world through the Fall of man (Gen. 3).

Genetic disease is not part of God's plan for this world. It is the result of the Fall. Christians can apply technology to fight these evils without fearing (as Dr. Reux does in *The*

Plague) that they will be fighting against God's will. This research could be used to cure genetic defects as a long-range goal and at the same time provide gene products in the short range in order to alleviate suffering.

A second type of research is in *the creation of new forms of life*. While most research at present is focused on curing genetic diseases, the trend is definitely moving from curing toward improving on nature through large-scale production of novel life forms. This is a drastic departure from previous scientific interventions into life. It is one thing to add one gene or a short gene complex to an organism and modify it slightly. It is quite another to create new forms of life. Conferring an insulin-production capability on a bacterial cell is very different from creating new creatures.

It is difficult to see how this type of genetic research could go unregulated. Freedom of scientific inquiry does not include the freedom to redesign a nature. Robert Sinsheimer has asked, "Would we wish to claim the right of individual scientists to be free to create novel self-perpetuating organisms likely to spread about the planet in an uncontrolled manner for better or worse?"[59] Christians should answer no.

Many scientists are already concerned about the wisdom of human manipulation of the environment. The major theme in the environmental movement is that mankind should not manipulate the environment from ignorance. But now there are many of these same scientists who are willing to manipulate life by creating new forms of life.

It should be noted that the manipulation of the environment is less serious in most cases than the manipulation of life. Ecosystems are marked by their resiliency and ability to "bounce back." This is not the case with genetic manipulation. Once a creature is altered in its genetic structure so that it can no longer exchange its genes with other creatures, it is fixed in its constitution.

The field of rDNA research is potentially more dangerous than that of any other scientific research. In most areas of scientific research, the potential hazards are short-lived and local. Except for the products of nuclear-power generation, no other by-product exists for extremely long periods of time. Even DDT, PCB, and HCB decompose naturally and relatively quickly. By contrast, the organisms created by rDNA research will always be in the environment.

Recombinant DNA research is also unique in the extent of its impact. The impact of releasing newly created forms of life in the environment will be felt outside of a local area. These new forms will spread to nearly every niche in the environment. This will not be only a local or a national problem but an international problem. Unhappy consequences following the introduction of the rabbit into Australia, the rat to Hawaii, and the gypsy moth into the continental United States should be sufficient reason for caution.

Such research is unique also because it can restructure all forms of life, including that of man himself. Leon Kass has said, "Engineering the engineer seems to differ in kind from engineering his engine."[60] It is likely that scientists will desire to redesign not only other life forms but man himself.

One day while taking a course in evolution at Yale, I was surprised to hear my professor say that he hoped scientists would soon create a human-primate hybrid. He suggested that a woman be impregnated with primate sperm. Later I learned that some of my female classmates had come forward to volunteer for the experiment if it was ever run.

Japanese scientists attempted to do such experiments a few years ago. (The lack of later publicity suggests that these experiments must have failed.) A Japanese actress volunteered to have sexual intercourse with an ape with forty-seven chromosomes (apes usually have forty-eight, humans forty-six in order to bear an ape-human hybrid.[61] Dr. Geoffrey Bourne of the Yerkes Primate Center has reported that Russians have contacted him and encouraged him to pursue such research.[62] He suspects that they may have already done such research and are looking for others to do the same so they can release their findings without suffering criticism from Western scientists.

Recombinant DNA techniques not only allow the minor genetic changes necessary for plant and animal breeding but also allow the large-scale mixing of genes of all living organisms (plant, animal and human). Christians must oppose such research. The dominion of humans over the created order does not include taking over the office of Creator. God created plants and animals as "kinds." While there was minor variability within these created "kinds," there are built-in barriers between kinds that we would do well to maintain. Redesigning creatures of any kind cannot be predicted in the same way that new elements on the periodic chart can be predicted for properties even before they are discovered. We simply do not know enough to begin redesigning life.

The creation of primate-human hybrids should not be allowed. This would violate the very essence of human dignity. We are created in God's image and such experiments should not be done.

Erwin Chargaff asked, "Have we the right to counteract, irreversibly, the evolutionary wisdom of millions of years, in order to satisfy the ambition and curiosity of a few scientists?"[63] His answer was no.

Since Christians believe that God is responsible for life on this planet, their answer should be even stronger. Life is not the result of millions of years of chance. It is here because God created it. Because humans are fallen, they are in need of redemption. Man's exercising dominion over the creation does not include creating new forms of life. One Christian periodical put it best when it argued that we are not called to "rewrite the fifth day of creation."[64]

In conclusion, we have seen that many of the scientific issues raised by opponents of rDNA research were successfully answered and many restrictions have been eased. Many legal questions remain, and Christians should be concerned that the present patent process has significantly changed the nature of scientific inquiry. Finally there are some legitimate ethical and theological concerns about this research that cause us to reject the idea that we can create new life forms. Recombinant DNA offers great promise in treating genetic disease but should not be used to confer the role of creator on mankind.

64773

VI

Human Cloning

In 1970 Alvin Toffler published his book *Future Shock* in which he quoted Nobel laureate Joshua Lederberg as saying that the cloning of a human would take place within fifteen years.[1] Also, that same year Paul Ramsey published his book *Fabricated Man* in which he devoted a lengthy chapter to the subject of human cloning.[2] But even with these advance predictions, many were caught off guard when David Rorvick claimed in his 1978 book *In His Image: The Cloning of a Man* that a man had recently been cloned.[3]

The claim of human cloning was most likely a hoax perpetrated to sell more books. Representative Paul Rogers called a special hearing of the Health and Environment Subcommittee to evaluate Rorvick's claim. Testimony at the hearing indicated that human cloning is still a distant possibility.

Beatrice Mintz of the Institute for Cancer Research (Fox Chase Cancer Center) in Philadelphia called the book by Rorvick "unquestionably a work of fiction."[4] Clement Markert, who refined the microsurgical techniques for cloning at Yale University, said, "I flatly disbelieve the claim."[5]

J. D. Broomhall at Oxford also has provided evidence that the Rorvick book was a hoax. According to his records, David Rorvick wrote to him for technical information about cloning five months after the clone was already supposedly born.[6]

Broomhall filed suit against the publishers and Rorvick,

Lincoln Christian College

charging them with defaming him by using his name in the book without permission.[7] Although the court dismissed the defamation charges, a recent ruling by U.S. District Judge John Fullam established the book as fiction.[8]

More recent research in this field has shown that cloning of humans may be possible in the future. Dr. Landrum Shettles reported in the *American Journal of Obstetrics and Gynecology* that he has cloned human embryos to the blastocyst stage.[9]

POTENTIAL BENEFITS OF CLONING

There are a number of benefits of cloning as a laboratory procedure and as a genetic technique. Horticulturists have cloned plants for centuries. In fact, the word *clone* comes from the Greek word *klon*, which means twig or offshoot. A form of cloning is taking a piece of a plant and rooting it in water. By taking the "eyes" of a potato, one can produce genetic clones of the original potato.

Cloning animals has been a more recent development. In 1952 Robert Briggs and Thomas King, working at the Institute for Cancer Research in Philadelphia, replaced the nuclei of unfertilized egg cells with the nuclei from blastula cells.[10] By 1961 J. B. Gurdon, working at Oxford University was able to clone tadpoles from adult frog cells (see Figure 3). He started by destroying the nuclei from unfertilized egg cells and then inserted the nuclei from intestinal cells of adult frogs. Some of the eggs began to divide and formed tadpoles that were genetically identical to the donor tadpole.[11]

The application of this technique to humans is still not within present research capabilities. There are differences between the size of frog eggs and human eggs and differences in their totipotency (ability to develop into a complete organism). Theoretically, scientists will be able to transfer this genetic technique to humans.

Cloning's greatest value is not as an alternative means of reproduction but rather as a powerful research tool in the laboratory. Its use as a research tool in developmental biology is very great. It will help in the study of nuclear differentiation so that we can understand how an embryonic cell becomes an eye cell or a toe cell. It will also be helpful in the study of immunology and organ rejection.[12]

Its use in medical research will also be of great benefit. Cloning can be used in the study of cancer. It also will be helpful in the study of aging. Dr. Leonard Hayflick, formerly at Stanford University, has cultured in-vitro cancerous cells that do not have the normal limited life span. With the use of serial nuclear transplantation (a cloning procedure), the aging process can be studied using these cells.[13]

Figure 3

CLONING

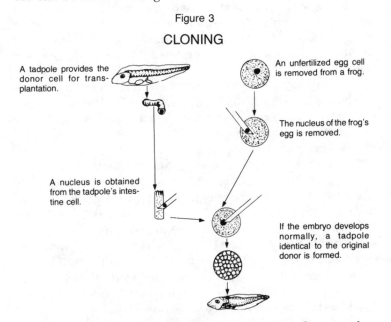

A tadpole provides the donor cell for transplantation.

An unfertilized egg cell is removed from a frog.

The nucleus of the frog's egg is removed.

A nucleus is obtained from the tadpole's intestine cell.

If the embryo develops normally, a tadpole identical to the original donor is formed.

Dr. Beatrice Mintz reported that her research group has recently used artificial manipulations to produce mice genetically endowed with Lesch-Nyham disease, thus providing animal models of human genetic disorders for study. Other research by her led to the production of mice from fertilized egg cells that were altered by replacing the cell nucleus with a nucleus of a cancer cell (a tetracarcinoma). Since the mouse developed normally, this suggested there is some internal process working to make the original cell noncancerous—a process that could have future application.[14]

Cloning also has application in the field of animal husbandry. Once a particular genetic strain of beef or pork is produced, cloning can be used to increase these genes in the population. It might also be used to retain the hybrid vigor of particular animals.

PROPOSED BENEFITS OF HUMAN CLONING

Some scientists feel that it would be beneficial to clone humans. Joshua Lederberg foresaw the time when the genetic fitness of the human race would be improved by the selective reproduction and cloning of individuals who best measured up to certain somatic, psychosociological, and genetic criteria.[15]

Dr. Bernard Davis has advocated that it may be advantageous to clone individuals "who might enormously enhance our culture" especially if they are "in fields such as mathematics or music, where major achievements are restricted to a few especially gifted people."[16]

Four hundred years ago something similar was attempted by using selective breeding. When Leonardo da Vinci died, his half brother Bartolommeo sought out a peasant girl who had the same traits as Leonardo's mother and had a child from her. The child, Pierino da Vinci, was raised in the same Tuscan countryside that nurtured Leonardo. At the age of twelve, he was taken to Florence to study under some of the well-known art teachers of that day. He showed great promise (a number of his works until recently were attributed to Michelangelo) until he died at the age of twenty-three.[17]

Many modern-day geneticists propose a similar experiment (but more elaborate) for major figures. J. B. S. Haldane argued that since we save the writings of famous people, we should also save their genotypes. He even suggested a state-supported retirement at age fifty-five so that each of these people would have time to raise and train his or her own clone.[18]

Many of these scenarios are what one scientist has called

"an interesting exercise in social science fiction."[19] They are full of naïve assumptions about our society that are unrealistic and basically unchristian. They also have a naturalistic view of human nature and assume that a cloned individual will be quite similar to the original donor, thus neglecting the contribution of our spiritual nature to personality development.

If we neglect some of these less-realistic scenarios, we can list five possible applications of human cloning suggested by scientists. The first is *biological immorality*. A given genotype has an extremely low probability of occurring on the earth more than once (with the exception of multiple births). Cloning provides a possible means of preserving a given genotype for more than one generation.

Some have suggested that this could be used retroactively to "resurrect" past genotypes. Dr. Elof Axel Carlson of the University of California at Los Angeles would like to see cloning used to revive dead personalities. He has suggested that there is perhaps enough material left of King Tutankhamen in his mummy case to produce a clone of him.[20] It is unlikely that this is possible, but the idea does raise issues that will be addressed later.

A second application would be to use cloning to *prevent genetic disease*. If both partners in a marriage were diagnosed as carrying a recessive gene for a genetic disease, they could clone one of them so that the child would only be a carrier and not be affected by the disease.

It is argued that not only would this allow them to prevent the disease, but it would allow them to pick both the sex and the physical characteristics of such children. Since it does not require sexual intercourse and genetic recombination, a couple could choose someone else to clone with complete assurance of the resultant characteristics and without the stigma that artificial insemination or other means might have.

There is at least one problem with this mode of reproduction if the mother is cloned and carries the baby. During pregnancy fetal cells often break loose and find their way to the mother's tissue. Normally, the immune system recognizes these tissues as "foreign" and rejects them. If the mother were carrying her clone, these tissues might not be rejected and could then spread through her body and become malignant. Thus there would be some danger of contracting cancer from a clone pregnancy.[21]

A third use would be in *social experimentation*. There is a continuing debate among scientists about the importance of heredity and environment. The study of twins provides helpful information concerning this question and scientists propose creating clones for further study.

While this may be promising, studies of twins have provided significant research material. One recent study involved twins who had been separated from each other for thirty-nine years.[22] These studies are sufficiently helpful to remove this suggested application of cloning from serious consideration.

Fourth, it has been suggested that cloning might improve *psychic communication*. It is felt that clones might experience a special intimate communication with other clones similar to that reported in twins. Some argue that it might be possible to breed these clones for a high degree of psychic compatibility and thus develop the powers of telepathy. Few scientists, however, feel that this would be a very likely prospect.

Cloning could also be used for *improved organ transplantation*. With colonies of clones in existence, freer exchange of organs would be possible without fear of organ rejection. Joshua Lederberg has even suggested that a clone of each person could be grown and kept in storage against the day when such an organ transplant would be necessary. This system would allow a "free exchange of organ transplants with no concern for graft rejection."[23]

SCIENTIFIC CONSIDERATIONS

When reproduction is accomplished asexually rather than by sexual means, there are some concerns. In the world of nature, asexual reproduction is a less efficient means of continuing the species since little genetic variability can be maintained. Any prokaryote that uses asexual reproduction cannot leave its offspring with enough variability. For this reason mass cloning is genetically undesirable.

Similarly, producing multiple clones of humans may be very hazardous. There are two major genetic problems with human cloning. First, cloning will increase the number of deleterious genes. The phenotype may be very desirable, but the genotype may contain many harmful genes that will be increased in the population by cloning.

Second, production of clones would increase the incidence of genetic disease in the population. Since clones of one person would be genetically identical to each other, their children would be half brothers and half sisters to each other. Unless there was extremely good record keeping and governmental regulation of marriages, the possibility of forming "incestous" relationships would be very great. People genetically similar to each other would be marrying each other, and recessive genes for genetic disorders would be expressed in the human gene pool.

There would also be the potential for clones to suffer reproductive difficulties. J. B. Gurdon found that cloned frogs showed an even higher incidence of sterility than those raised in the laboratory, and the sterility incidence of those raised in the laboratory, in turn, was higher than for those found in the wild.[24] This may also occur in human clones.

This leads us to think of additional concerns about the safety of using artificial reproductive and genetic techniques. Many of the concerns raised in chapter 4 about in-vitro fertilization are relevant here. Physical and mental disorders may result from the artificial manipulation of eggs. Most cloning experiments have resulted in many abnormalities, and there is cause for concern over other possible abnormalities that would be less obvious.

SOCIAL AND LEGAL CONSIDERATIONS

The possibility of producing abnormal people through cloning raises grave concerns. These are similar to the issues raised concerning in-vitro fertilization. Cloning humans would most likely create many social and legal difficulties. One law journal has noted:

> Protection of the interests of society as a whole must be considered. Any errors in cloning will be irreversible. The government might be confronted with the problem of arbitrarily classifying the progeny, which are to be killed, or as humanoids, who will be permitted to live at government expense as a reminder of the imperfect operation of cloning techniques.[25]

The best protection would be never to use cloning techniques on human subjects at all.

Cloning would also raise the social concern of self-image. Clones would suffer from this more than twins. Twins have to forge their identity in the presence of one genetic copy. Clones would most likely have many genetic copies in existence. They would also have greater difficulty in self-identity due to the fact that they would be following the path of an already-existing (identical) donor. Imagine the pressure on the clone of an Einstein to measure up to the status of the original. There might also be a loss of a sense of self-worth because the clone was "manufactured" rather than produced by normal means.

The effect of cloning on society would also be of concern. Increasing the variety of reproductive modes may begin to erode rather than enhance personal freedom. Laurence Tribe, professor of constitutional law at Harvard Law School, has said:

> If we were to have a menu of alternative ways by which to propogate the species—to begin with, sexual reproduction, the random combination of genotypes, the ordinary method today; but in addition to that, the deliberate replication of a chosen genotype; and then finally the design of a new genotype—I think it would not be very long before centralized decision-making, rather than personal choice, would determine which of those methods you or I could pick, what we could pick from the menu.[26]

Some of the same legal questions raised in previous chapters are also relevant here. Would it be possible to patent cloning techniques? Would clones have to pay royalties to the genetic donors? Today we pay authors royalties for copies of their writings. Would the courts rule that a person deserves some royalty for allowing copies of himself to be made?

Some of the same social questions raised previously would also be appropriate. In particular the question of human parenthood is important. In a normal sexual union the progeny are the product of two sets of genetic materials. Cloning requires only one parent. The other parent is not a participant or even genetically related to the child. The exclusion of one parent recalls to mind questions concerning artificial insemination by a donor (AID), but the difference is that the cloned child would be *completely* the offspring of one parent.

The effect might be quite profound. A child born of normal sexual relations shares both heredity and environment with his two parents. When one parent "sits on the sidelines," a disruptive effect is possible. A cloned son would truly be his "dad's boy." Again, we see biological procreation evolving into mechanical and artificial reproduction.

ETHICAL CONSIDERATIONS

Cloning human subjects will begin the slide down a slippery slope to unwanted consequences. While there are many beneficial applications of cloning animals in the laboratory, the application to humans will not be desirable.

Many scientists are looking forward to a time when cloning on humans is used extensively. Dr. James Bonner has said:

> There is nothing to prevent us from taking two body cells from (a) donor and growing two identical twins.... As a matter of fact, there is nothing to prevent us from taking a thousand. We could grow any desired number of genetically identical people from individuals who have desirable characteristics.[27]

Of even greater concern is the suggestion that cloning would be used to provide organ parts for transplantation. If there are multiple clones made from the same donor, this would allow free organ exchange but would result in lower genetic diversity. Therefore, scientists have suggested that a clone could be raised in an incubation unit. If a heart were needed, a clone heart would be used.

The humanity of the clone is never considered in such discussions. Even though clones would be formed by a different mode of reproduction than is normal, they would most likely have souls and be God's image-bearers just as we are. To deny the humanity of the clone would be no different from denying the humanity of any other human being. This scenario of cloned parts is a grave one indeed.

Other ethical questions surround the possibility of increasing genetic disease through cloning. The only solutions would be to prevent clones from reproducing or to keep extensive records of genetic types that could be checked before a descendant of a clone would be allowed to reproduce. Neither choice is very appealing.

The possibility that many human clones may turn out abnormally also raises many ethical questions. Who would be responsible? How do we obtain informed consent? These are questions that have been asked of other genetic techniques and that haunt these idealistic cloning scenarios.

THEOLOGICAL CONSIDERATIONS

The fundamental theological issue of cloning is the sanctity of life. The potential for loss of life and genetic abnormality is very great with cloning experiments. Cloning animals and plants provides promise in some areas, but cloning human beings is a fundamentally different proposition. We are created in God's image (Gen. 1:27) and must be treated differently from the way other creatures in the world are treated.

Cloning also raises a question about what that image may be. There may be some wisdom in maintaining human diversity. With the exception of twins and other multiple births, each person has a unique genotype. There is unity in the human family but also diversity. Each person contributes to both the unity and the diversity of humanity. This is perhaps best expressed in the following Jewish Midrash:

> For a man stamps many coins in one mold and they are all alike; but the King who is king over all kings, the Holy One blessed be he, stamped every man in the mold of the first man, yet not one of them resembles his fellow.[28]

It is most likely that a clone would also be created in the image of God. Though the Bible does not definitively speak on this subject and theologians rarely address such questions, there is some reason to believe that clones might indeed have souls.

The greater question is whether their humanity would be redefined. On the basis of our previous history, it is reasonable to conclude that they would be used (for spare parts) and abused. The sanctity of human life is no longer accepted by much of our society. The humanity of a clone would be easily disregarded.

In conclusion, Christians must reject the future plans to clone humans. There are many scientific, social, legal, ethical and theological concerns and few significant benefits. While cloning has many laboratory applications, it should not be used as an alternative mode of reproduction for humans.

Genetics
and the
Christian

VII
Christian Responsibility

We are in the genetic age. Just as the nuclear age has brought both good and evil, so we can likely expect the genetic revolution to bear both good and bad fruits. If Christians are going to be salt and light to the culture (Matt. 5:13–16), then Christian responsibility must be exercised in this field of genetic engineering. Any technology is only as good or as bad as the world view behind it. A proper view of the world, society, and the human race is the best preventive measure against the abuse of a technology. Christians must be willing to think, speak, and act from a Christian base in this area of genetic technology.

GENETICS AND HUMAN LIFE

The commercialization that has followed advances in genetic engineering and artificial reproduction is very disturbing. Sperm banks pay $20 to men who offer their ejaculations and then sell the frozen sperm to subscribing physicians for $35. Genetic counseling firms offer sex selection to interested parents. Gene-splicing businesses plan a marketing strategy to maximize their investments in a patented microbe that produces human insulin. Surrogate mothers are sought by surrogate parenting groups for fees of $10,000 to $20,000. IVF clinics are offering in-vitro fertilization for a fee of about $4,000. Everything about reproduction is being commercialized!

⌞ One cannot help but wonder if our humanity is being diminished and commercialized in the process. Men are selling their sperm and women are renting their wombs. Each new advance brings us another step toward further commercialization. With this commercial market for sex products, we begin to wonder if there is really a fundamental distinction between selling one's body for sex (prostitution) and selling its sex products (sperm, womb for rent). ⌝

Are we moving into a period in our history where we will be judged on the basis of our genetic structure? The rise of the field of sociobiology seems to suggest this. Edward Wilson at Harvard University attempts to explain most of our behavior in the light of genetics. According to Wilson, we can account for human behavior and human culture on the basis of our genetic structure. Our actions are tied to our need to reproduce our genes. We are DNA's way of making more DNA. Those with good genes will have their genes reproduced, and those with bad genes will not and should not.

⌞ We should be concerned about a future that may move us into the genetic age with a form of genetic racism. In the past if you were producing children with genes unacceptable to society, that was unfortunate. Now that we have the possibility of eliminating those genes, will it become a crime to produce a child with a genetic defect?

The specter of "quality of life" standards is not a welcome one. Yet this seems to be what the future holds unless there is a radical change in our society's view of human life. Already, abortions for "wrong" genes have been allowed (genetic disease, "wrong" sex gene). How far are we from demanding genetic quality standards? ⌝

GENETICS AND THE FUTURE

Many commentators on genetic engineering have had a relatively pessimistic view of the future. Paul Ramsey, for example, fears the dehumanizing effects of this technology and is concerned that we will continue taking steps and slide down a slippery slope to unintended and immoral consequences.

The ultimate goal of reducing suffering from genetic disease is very noble. We can remove many of the physical effects of the Fall (Gen. 3) by applying genetic technology. For example, by using recombinant DNA research, it may be possible eventually to repair genetic defects in our gametes or in a developing embryo. Christians can and should support therapeutic attempts to treat, cure, and eliminate genetic diseases.

There is, however, another goal that many scientists have. Their goal is to use genetic technology to improve (actually re-create) the human race. They see these new tools as a means to perfect humanity.

Throughout history people have sought to improve the human race. Schemes for such improvement are as ancient as Plato's *Republic* and as modern as B. F. Skinner's *Walden Two* or E. O. Wilson's *Sociobiology*. But now scientists have available to them the tools to accomplish their dreams of eugenics.

Francis Galton, a cousin of Charles Darwin, founded modern eugenics in 1883. Influenced by the writings of the Enlightenment and his cousin's popular writings on evolution, Galton published his own book entitled *Hereditary Genius*, which outlined the application of eugenics to society.

Galton set out to eliminate the unfit from British society. He believed that curvature of the spine, club feet, and the high-arched palate were marks of criminality passed on from parent to child. He also thought drunkenness and epilepsy were signs of degeneration. Once this "bad seed" was evident in a family, he said, the tainted should be kept from breeding.

Galton asked for laws to prevent reproduction by epileptics, feeble-minded persons, convicted criminals, and paupers. He favored giving certificates of genetic merit to "healthy" young men and women. His successes were then multiplied by the social Darwinists, who rallied around the cry of "survival of the fittest." This new theory was used to justify everything from monopolistic business practices to racism.

In this century eugenics was promoted by the renowned American geneticist Herman Muller. He once remarked that we are all fellow mutants together and that the only way to avoid the reproduction of the ordinary was to take purposeful control over human reproduction.

Muller argued that programs of planned eugenics would provide the opportunity to guide human evolution. By instituting such programs, he argued, society would make "unlimited progress in the genetic constitution of man, to match and re-inforce his cultural progress and, reciprocally, to be re-inforced by it, in a perhaps never-ending succession."[1]

Shortly before World War II many prominent scientists proposed a "geneticist's manifesto" that articulated their desires to apply human eugenic theory. It called for the genetic improvement of mankind through the legalization, development, and dissemination of birth-control systems (both positive and negative) that would be put into effect at all stages of the reproductive process. It also called for those with a better genetic constitution to produce more offspring than the rest of the population.[2]

With the outbreak of World War II and the later revelations of Nazi breeding experiments, people were repulsed by them and the eugenics movement encountered a major setback. But now there seems to be a renaissance in thinking in this area. The naturalistic, evolutionary world view is making a significant impact on our thinking today. Many are calling for the institution of a modern form of eugenics.

Joseph Fletcher, author of *Situation Ethics* and *The Ethics of Genetic Control*, for example, has argued that the government must establish human quality-control standards. At a 1979 Symposium on Genetics and Law in Boston, he said that "coercive or compulsory control is justified" in cases where carriers of genetic disease do not abstain voluntarily from having children.[3]

Fletcher has further suggested that a system of genetic licensing be implemented. He noted, "We will soon see how absurd it is that we have to be issued drivers' licenses to operate an automobile, while we are free to go on producing children of any kind we want or happen to conceive."[4]

Nobel prizewinner Linus Pauling has suggested that

people diagnosed as carriers of genetic diseases (all of us carry from three to eight genetic defects) be conspicuously marked:

> I have suggested that there should be tattooed on the forehead of every young person a symbol showing possession of the sickle-cell gene or whatever other similar gene, such as the gene for phenylketonuria, that he has been found to possess in a single dose. If this were done, two young people carrying the same seriously defective gene in a single dose would recognize this situation at first sight, and would refrain from falling in love with one another. It is my opinion that legislation along this line, compulsory testing for defective genes before marriage, and some form of public or semi-public display of this possession, should be adopted.[5]

One cannot help but be fearful of the striking similarity between Pauling's desire for tatoos on the forehead and the biblical description of the mark of the beast in Revelation.

The possible satanic application of genetic technology to humans cannot be ruled out. Carl Rogers has noted that we possess the ability to completely manipulate human behavior:

> We can choose to use our growing knowledge to enslave people in ways never dreamed of before, depersonalizing them, controlling them by means so carefully selected that they will perhaps never be aware of their loss of personhood.[6]

We are moving quickly toward a society that will control human nature. Man will become the controller and manipulator of brain and body. Charles Frankel, in his article on the "Specter of Eugenics," expressed the following fear:

> Biomedicine, we are told, is the harbinger, or portent, of the day when man will be able to say of himself, meaning it entirely, that, at last, he is his own creation, and has got the weight of that other Creation off his back.[7]

Many non-Christian scientists seek human genetic engineering because they have rebelled against their Creator. Their agenda extends far beyond what Christians feel is appropriate.

Therefore, Christians must work to control the application of genetic engineering in the future.

One important way to control this technology is to have governmental limits on this technology. Christians usually have been effective in identifying the problem and fairly good at evaluating the ethical issues but very weak in prescribing and implementing remedies.

GENETICS AND POLITICS

Many have felt that genetic policy should be decided by professionals in the field (scientists and bureaucrats involved in federal funding of genetics). There is, however, grave danger in letting elite professional groups decide for the masses. Scientists involved in rDNA research may know a great deal about their field but most of them are probably not competent to consider the social, ethical, or theological implications of their research. Physicians studying reproductive biology may have a handle on the scientific considerations of AIH, AID, and IVF, but most of them are unlikely to be in a position to decide whether federal funding or control should be implemented for modes of artificial reproduction.

As much as possible, questions concerning genetics must be decided by the public through their elected representatives. There should be greater emphasis placed on the discussion and evaluation of these issues. Congressional hearings on genetics legislation and regional public hearings such as those held by the Ethics Advisory Board of the Department of Health, Education, and Welfare would be examples of responsible governmental action that has been taken and should be continued.

Another suggestion would be to implement a science court. Dr. Arthur Kantrowitz first suggested this in the 1960s, and the idea has gained support from many science organizations.[8] Recently, Professor Etzioni at George Washington University suggested that such a court be convened to judge the merits of economist Arthur Laffer's proposals before they are implemented by the Reagan administration.

A science court for genetics would be similar to the one

that was convened by the Cambridge City Council concerning rDNA research at Harvard. It would act as a bridge between the science community and policy-making bodies. Judges would be drawn from the ranks of scientists and laymen in an effort to form an impartial board. They would have the power to supoena witnesses on both sides of an issue and could arbitrate between competing claims. This would provide an important fact-finding body for legislators and governmental officials.

Much of the information necessary to make a judgment on a particular point of genetics or reproductive biology would be scientific, social, or legal. Official decisions must, however, be in harmony with biblical absolutes. As Christians we must not shrink away from the policy arena but must rely on our biblical heritage for help in cutting through some of the competing claims made for each genetic advance.

Christians must have a discerning attitude toward the potential problems that genetic policies might create. All too often people have tended to laud the benefits of genetic engineering and neglect the costs. Given the fact that we live in a fallen world (Gen. 3) and are sinful people (Rom. 3:23), we must guard against the proliferation of evil. We cannot expect people who do not accept the notion of human sinfulness and lack Christian commitment to protect society from disaster. Christians must participate in the policy-making process.

If Christians are not involved, then disaster is imminent. People have the technological ability to play God but are sinful and cannot act like Him. The witness of the humility of Christians before God will remind others that He is worthy of worship and obedience. We must seek His wisdom as we travel down the dangerous path of the genetic age.

Notes

Chapter One

[1]"Genetic Engineering: Reprise," *Journal of the American Medical Association* 220 (1972): 1356.

[2]Bruce K. Waltke, "Reflections from the Old Testament on Abortion," *Journal of the Evangelical Theological Society* 19 (Winter 1976): 3–13.

[3]Graham A. D. Scott, "Abortion and the Incarnation," *Journal of the Evangelical Theological Society* 17 (Winter 1974): 29–44.

[4]Lori B. Andrews, "Embryo Technology," *Parents*, May 1981, pp. 63–64.

[5]Ibid., p. 64.

[6]Edward L. Tatum, "Perspective from Physiological Genetics," in *The Control of Human Heredity and Evolution*, ed. T. M. Sonneborn (New York: Macmillan, 1965), p. 22.

Chapter Two

[1]M. Curie-Cohen, M. Luttrell, and S. Shapiro, "Current Practice of Artificial Insemination by Donor in the United States," *New England Journal of Medicine* 300 (1979): 585–90.

[2]Ibid.

[3]William G. Johnson, Robin C. Schwartz, and Abe M. Chutorian, "Artificial Insemination by Donors: The Need for Genetic Screening," *New England Journal of Medicine* 304 (1981): 755–57.

[4]Glanville L. Williams, *The Sanctity of Life and the Criminal Law* (London: Faber and Faber, n.d.), p. 118.

[5]"Dr. Andrew E. Good, of the Evanston Hospital," *Houston Chronicle*, 20 March 1980, sec. 1, p. 6.

[6]John B. Gurdon, "Some Legal Considerations," *Soundings* 54 (Fall 1971): 310–11.

[7]Curie-Cohen, Luttrell, and Shapiro, "Current Practice of Artificial Insemination."

[8]"Lesbian Story Stuns Britain," *Dallas Times Herald*, 7 January 1981, p. A-5.

[9]"Surrogate Birth Gives Baby Boy to Transsexual," *Dallas Times Herald*, 17 January 1981, p. A-7.

[10]Karl Ostrom, "Psychological Considerations in Evaluating AID," *Soundings* 54 (Fall 1971): 325–30.

[11]Aphrodite Clamar quoted by Carin Rubenstein in "Little Known Hazards of AID: Disease, Inbreeding, Guilt," *Psychology Today*, May 1980, p. 23.

[12]Elizabeth Bibb, "Childbearing Surrogates Not 'Kooks,'" *Dallas Times Herald*, 3 February 1981, pp. B-1, 7.

[13]"Surrogate Baby Deal Completed," *Dallas Times Herald*, 21 April 1981, p. A-3.

[14]Elizabeth Bibb, "Bearing Baby for Love or Money?" *Dallas Times Herald*, 2 February 1981, pp. B-1, 7.

[15]"Woman Offers to Bear Child for $15,000 Fee," *Dallas Times Herald*, 21 September 1981.

[16]Elaine Markoutsas, "Women Who Have Babies for Other Women," *Reader's Digest*, August 1981, p. 72.

[17]"Surrogate Parenting Under Fire in Kentucky," *Dallas Times Herald*, 22 January, 1981.

[18]"Surrogate Mother Wins Custody," *Dallas Times Herald*, 6 June 1981.

[19]"Surrogate Mother Wants to Keep Child She Bore," *Dallas Times Herald*, 22 March 1981, p. A-9.

[20]"Reformed Mother," *Parade*, 11 February 1979, p. 14

[21]"Breeding for IQ's—A Plan Under Fire," *U. S. News and World Report*, 24 March 1980, p. 49.

[22]"Sperm Banker Robert Graham Wants the Brave New World to Be Sired by Nobel Prize Winners," *People*, 17 March 1980, pp. 34–36.

[23]"People," *Time*, 25 November 1974, p. 67.

[24]Vance Packard, *The People Shapers* (New York: Bantam, 1977), p. 266.

[25]Anne Taylor Fleming, "A New Kind of Baby Boom," *San Francisco Chronicle*, 3 August 1980, p. 5.

[26]Elizabeth Bibb, "Bearing Baby for Love or Money?" *Dallas Times Herald*, 2 February 1981, p. B-7.

Chapter Three

[1]Pat K. Lynch, "Women: The Next Endangered Species?" *Madamoiselle*, May 1977, p. 32.

[2]David Rorvick with Landrum B. Shettles, *Your Baby's Sex: Now You Can Choose* (New York: Bantam, 1970).

[3]R. J. Ericsson, C. N. Langevin, and M. Nishino, "Isolation of Fractions Rich in Human Y Sperm," *Nature*, 14 December 1973, p. 422.

[4]"Sex Control Specialists, *"Dallas Morning News*, 7 September 1978, p. 1AA.

[5]David Rorvick, *Brave New Baby* (New York: Pocket Books, 1971), p. 64.

[6]Ibid., p. 67.

[7]*National Fertility Study* (1970), coordinated by Norman B. Ryder and Charles F. Westoff of the Office of Population Research, Princeton University.

[8]Charles F. Westoff and Ronald R. Rindfuss, "Sex Preselection in the United States: Some Implications," *Science*, 10 May 1974, pp. 633–36.

[9]Lynch, "Women," p. 34.

[10]Amitai Etzioni, "Sex Control, Science and Society," in *Heredity and Society*, ed. Adela S. Baer (New York: MacMillan, 1973), pp. 238–39.

[11]Westoff and Rindfuss, "Sex Preselection," p. 636.

[12]Marc Lappe, "Choosing the Sex of Our Children," *The Hastings Center Report*, Institute of Society, Ethics and the Life Sciences, February 1974, p. 3.

Chapter Four

[1]Leon R. Kass, "Babies by Means of In-Vitro Fertilization: Unethical Experiments on the Unborn?" *New England Journal of Medicine* 285 (18 November 1971): 1174–79; Paul Ramsey, "Shall We Reproduce? *Journal of the American Medical Association* 22 (5 and 12 June 1972): 1346–50, 1480–85.

[2]Joseph Fletcher, *The Ethics of Genetic Control* (Garden City, N.Y.: Doubleday, 1974); R. G. Edwards, "Fertilization of Human Eggs In-Vitro: Morals, Ethics, and the Law," *Quarterly Review of Biology* 49 (1974): 3–26; Robert G. Edwards and David J. Sharpe, "Social Values and Research in Human Embryology," *Nature* 231 (14 May 1971): 87–91.

[3]Protection of Human Subjects: Policies and Procedures, *Federal Register* 39 (23 August 1974): 30, 650.

[4]Ethics Advisory Board, *Reports and Conclusions: HEW Support of Research Involving In-Vitro Fertilization and Embryo Transfer* (Washington, D.C.: Government Printing Office, 1979).

[5]World Medical Association, "Declaration of Helsinki," *Medical Journal of Australia* 1(14 February 1976): 206; Leon Kass, "Making Babies—The New Biology and the 'Old' Morality," *Public Interest* 26 (Winter 1972): 28; Marc Lappe, "Risk-Taking for the Unborn," *Hastings Center Report* 2 (February 1972): 3; Benjamin Brackett, "Comment on Human Fertilization In-Vitro"—statement before the Subcommittee on Health and the Environment, House Committee on Interstate and Foreign Commerce, 4 August 1978.

[6]Ethics Advisory Board, *Report and Conclusions*, p. 5.

[7]"First Test Tube Calf Extremely Happy," *Dallas Times Herald*, 11 June 1981, p. 6-A.

[8]Lori B. Andrews, "Embryo Technology," *Parents*, May 1981, p. 66.

[9]Luigi Mastroianni quoted by V. Cohn in "Lab Growth of Human Embryo Raises Doubt of Normality," *The Washington Post*, 21 March 1971.

[10]Steven Hecht and Marc Lappe, "Moratorium on Human Zygote Implantation," *New England Journal of Medicine* (28 September 1978): 672.

[11]"Biologist: Test Tube Babies May Present Risk," *Houston Chronicle*, 25 January 1980, sec. 3, p. 9.

[12]Kass, "Babies," *New England Journal*, p. 1176.

[13]"Test Tube Results," *Christianity Today*, 18 August 1978, p. 36.

[14]"That Baby Again," *Time*, 19 February 1979, p. 82.

[15]Andrews, "Embryo Technology," *Parents*, pp. 63–70.

[16]J. G. Boue and A. Boue, "Increased Frequency of Chromosomal Abnormalities in Abortions After Induced Ovulation," *Lancet* 1 (1973): 679.

[17]L. R. Fraser et al., "Increased Incidence of Triploidy in Embryos Derived from Mouse Eggs Fertilized In-Vitro," *Nature* 260 (1976): 39–40.

[18]M. Ahlgren, "Sperm Transport to and Survival in the Human Fallopian Tube," *Gynecological Investigations* 6 (1975): 200–14.

[19]Jean Seligmann, "Test Tube Clinic," *Newsweek*, 5 March 1979, p. 102.

[20]B. D. Cohen, "New Medical Technique: Hope for Childless Couples," *The Washington Post*, 18 February 1980, p. C6.

[21]"Va. Blue Cross Bars Payments of Test-Tube Baby Program Fees," *The Washington Post*, 21 February 1980, p. A13.

[22]Glenn Frankel, "Test-Tube Baby Clinic," *The Washington Post*, 9 January 1980, p. C1.

[23]Glenn Frankel, "Test-Tube Baby Clinic Wins Another Round," *The Washington Post*, 13 February 1980, p. C1.

[24]"Test-Tube Baby Clinic Opponents Give Up Efforts for Court Order," *The Miami Herald*, 4 January 1981, p. 6A.

[25]Ethics Advisory Board, *Report and Conclusions*, p. 72.

[26]Ibid., p. 74.

[27]Jerold K. Footlick, "Test-Tube Bereavement," *Newsweek*, 31 July 1978, p. 70.

[28]*Del Zio v. Presbyterian Hospital*, 1974 Civ. 3588 (S. D. N. Y. 1978).

[29]"Hired Stud," *Parade*, 25 June 1978, p. 13.

[30]"Investigation Urged on Test Tube Babies," *Dallas Times Herald*, 29 October 1981 (no p. no.).

[31]Andrews, "Embryo Technology," *Parents*, p. 67.

[32]Ibid., p. 69.

[33]Jon Anderson, "Will Pregnant Men Be in a Pickle or Just Craving One?" *Dallas Times Herald*, 12 August 1981, p. 6-F.

[34]John Kelley, "Lifeshock: How Science Is Changing the Human Race," *Mademoiselle*, November 1981, p. 159.

[35]Leon Kass, "Babies by Means of In-Vitro Fertilization: Unethical Experiments on the Unborn?" *New England Journal of Medicine* 285 (18 November 1971): 1178.

[36]Bentley Glass, "Science: Endless Horizon or Golden Age?" *Science* 171 (8 January 1971): 28.

[37]Andrews, "Embryo Technology, *Parents*, p. 67.

[38]"Lesbian Birth Story Stuns Britain," *Dallas Times Herald*, 7 January 1978, p. A-5.

[39]"Sixteen Months After His Death, Cartoonist Kim Casali Gives Birth to Her Husband's Miracle Baby," *People*, n.d.

[40]Leon R. Kass, "Making Babies Revisited," *Public Interest* 54 (Winter 1979): 49.

[41]B. Rensberger, "From the Day of Deposit: A Lien on the Future," *The New York Times*, 22 August 1971.

[42]Robert Locke, "Moral Furor," *Herald-Leader*, 1 March 1980, pp. 1, 10.

[43]Kelly, "Lifeshock," p. 161.

[44]Kass, "Making Babies," pp. 29–30.

[45]Paul Ramsey, *Fabricated Man* (New Haven, Conn.: Yale University Press, 1970).

[46]C. S. Lewis, *The Abolition of Man* (New York: MacMillan, 1947), pp. 69–71.

[47]Richard Cohen, "Test-tube Babies: Why Add to a Surplus?" *The Washington Post*, 3 February 1980, p. B-1.

[48]Gina Bari Kolata, "Infertility: Promising New Treatments," *Science* 202 (October 1978): 201–3.

[49]Ramsey, "Shall We Reproduce?" p. 1482.

[50]Kass, "Making Babies," p. 54.

[51]André Hellegers and Richard McCormick, "Unanswered Questions on Test Tube Life," *America*, 139 (19 August 1978): 76.

[52]Allen Verhey, "Test Tube Babies: Two Responses," *The Reformed Journal*, September 1978, p. 16.

[53]Ramsey, "Shall We Reproduce?" p. 1487.

Chapter Five

[1]Ethan Singer quoted by Nicholas Wade in "Gene Splicing: Congress Starts Framing Law for Research," *Science* 196 (1 April 1977): 39.

[2]Michael Crichton, *The Andromeda Strain* (New York: Dell, 1969).

[3]A summary of the N.A.S. report can be found in the article "Asilomar Conference on DNA Recombinant Molecules," *Nature* 255 (5 June 1975): 442–44.

[4]Erwin Chargaff, "A Slap at the Bishops of Asilomar," *Science* 190 (10 October 1975): 135.

[5]Decision of the director of National Institute of Health to Release Guidelines for Research on Recombinant DNA Molecules, *Federal Register* 41 (1976): 27902–03.

[6]Nicholas Wade, "Recombinant DNA: NIH Group Stirs Storm by Drafting Laxer Rules," *Science* 190 (1975): 769.

[7]Decision of the director, pp. 27, 902–43.

[8]"The DNA Furor: Tinkering With Life," *Time*, 18 April 1977, p. 45.

[9]Shing Chang and Stanley N. Cohen, "In-Vivo Site-Specific Genetic Recombination Promoted by the *Eco*-RI Restriction Endonuclease," *Proc. Natl. Acad. Sci. USA* 74 (November 1977): 4811–15.

[10]Nicholas Wade, "Major Relaxation in DNA Rules," *Science* 205 (21 September 1979): 1238.

[11]"Virus Cloning Accident," *San Francisco Chronicle*, 8 August 1980, pp. 1, 28.

[12]Peter Gwynne, "DNA and Insulin," *Newsweek*, 6 June 1977, p. 74.

[13]"Creating Insulin," *Time*, 18 September 1978, p. 102.

[14]"Genetic Insulin Tested," *Dallas Times Herald*, 13 February 1981, p. A-7.

[15]Patrick Young, "Gene Splicing: The Future of Genetic Engineering Is Now," *Dallas Times Herald*, 11 February 1981, p. 4, n.d.

[16]"Genetic Engineering Employed to Create New Influenza Vaccine," *Dallas Times Herald*, 17 October 1981, p. A-33.

[17]Matt Clark et al., "The Miracles of Spliced Genes," *Newsweek*, 17 March 1980, p. 64.

[18]"Synthetic Interferon Tested on Humans," *San Francisco Chronicle*, 16 January 1981, p. 6.

[19]Bernard David, "Potential Benefits Are Large, Protective Methods Make Risks Small," *Chemical and Engineering News*, 30 May 1977, p. 27.

[20]Richard Hutton, *Bio-Revolution: DNA and the Ethics of Man-Made Life* (New York: New American Library, 1978), p. 120.

[21]Claudia Wallis, "Tampering With Beans and Genes," *Time*, 19 October 1981, pp. 90–91.

[22]"New Life Forms: A Clear Road Ahead?" *U. S. News and World Report*, 30 June 1980, p. 34.

[23]Art Kaufman, "Research Scientists Investigate Non-medical Genetic Advances," *Dallas Times Herald*, 11 February 1981, p. 4ND.

[24]"Bacteria May Eat Toxins," *Dallas Times Herald*, 22 June 1980, p. A-23.

[25]Lee Dembart, "Biologists Transplant Human Gene Into Mouse," *Dallas Times Herald*, 12 December 1981, p. 42-A.

[26]Sharon Begley, "Curing Disease With Genes," *Newsweek*, 21 April 1980, p. 80.

[27]"Furtive First: Genetic Jump From Mice to Man," *Time*, 20 October 1980, p. 57.

[28]David Perlman, "Biochemist's Attack on Gene-Splicers," *San Francisco Chronicle*, 3 December 1980, p. 6.

[29]W. French Anderson and John C. Fletcher, "Gene Therapy in Human Beings: When Is It Ethical to Begin?" and Karen E. Mercola and Martin J. Cline, "The Potentials of Inserting New Genetic Information," *New England Journal of Medicine* 303 (27 November 1980): 1293–1300.

[30]Nicholas Wade, "Recombinant DNA: The Last Look Before the Leap," *Science* 192 (16 April 1976): 236.

[31]Bernard Davis, "Evolution, Epidemiology, and Recombinant DNA," *Science* 193 (1976): 442.

[32]Nicholas Wade, "Genetics: Conference Sets Strict Controls to Replace Moratorium," *Science* 187 (14 March 1975): 932.

[33]Nicholas Wade, "New Rulebook for Gene Splicers Faces One More Test," *Science* 201 (18 August 1978): 601.

[34]Wade, "Genetics: Conference," p. 932.

[35]Hutton, *Bio-Revolution*, pp. 155–56.

[36]Sheldon Krimsky, "Regulation No Threat to Free Scientific Inquiry," *Chemical and Engineering News*, 30 May 1977, p. 40.

[37]Bernard David, "Pending Law Could Seriously Impede Scientific Research," *Chemical and Engineering News*, 30 May 1977, p. 42.

[38]George Wald, "The Case Against Genetic Engineering," *The Sciences* 16 (May 1976): 10.

[39]Gene Bylinsky, "Industry Is Finding More Jobs for Microbes," *Fortune*, February 1974, pp. 96–102.

[40]John S. DeMott, "Test-tube Life: Reg. U.S. Pat. Off," *Time*, 30 June 1980, p. 52.

[41]Nicholas Wade," Supreme Court to Say If Life Patentable," *Science* 206 (9 November 1976): 664.

[42]DeMott, "Test-tube Life," p. 52.

[43]Thomas O'Toole, "Patenting Life Forms: A New Science Frontier," *Dallas Times Herald*, 22 June 1980, p. A-23.

[44]"High Court Ruling May Spur Genetic Research," *Dallas Times Herald*, 17 June 1980, p. 4-A.

[45]Testimony by Ronald Cape at the Subcommittee on Science, Technology, and Space of the Senate Committee on Commerce, Science, and Transportation, *Hearings*, 10 November 1977, p. 338.

[46]Spyros Andreopoulos, "Gene Cloning by Press Conference," *New England Journal of Medicine* 302 (27 March 1980): 743–45.

[47]David Pauly, "Waiting for Genentech," *Newsweek*, 20 October 1980, p. 72.

[48]"Investors Dream of Genes," *Time*, 20 October 1980, p. 72.

[49]DeMott, "Test-tube Life," p. 53.

[50]Ibid.

[51]Charles Petit, "New Gene-Splice Patent Granted," *San Francisco Chronicle*, 4 December 1980, p. 2.

[52]"Memo Spells Out Technology Transfer Policies," *Harvard Gazette*, 31 October 1980, pp. 4-5.

[53]DeMott, "Test-tube Life," p. 53.

[54]Ibid.

[55]Erwin Chargaff's remarks made before the National Academy of Sciences Forum on Recombinant DNA Research, Washington, D.C., 7 March 1977.

[56]Testimony by Ethan Singer before the Subcommittee on Health and the Environment of the House Committee on Interstate and Foreign Commerce, *Hearings*, 15 March 1977, p. 79.

[57]Julian Huxley quoted in Joseph Fletcher's book *The Ethics of Genetic Control* (Garden City, N.Y.: Anchor, 1974), p. 8.

[58]Nicholas Wade, "Recombinant DNA: The Last Look Before the Leap," *Science* 192 (16 April 1976): 237.

[59]Liebe F. Cavalieri, "New Strains of Life—or Death," *New York Times Magazine*, 22 August 1976, p. 60.

[60]Leon R. Kass, "The New Biology: What Price Relieving Man's Estate?" *Science* 174 (19 November 1971): 779.

[61]"Ape to Mate with 19-year Old Woman," article listed in *Bible-Science Newsletter*, October 1977, p. 6.

[62]"Test-tube Apeman Can Now be Made," *Modern People*, 26 December 1976.

[63]Wald, "The Case Against Genetic Engineering," pp. 6-11.

[64]Nancy McCann, "The DNA Maelstrom: Science and Industry Rewrite the Fifth Day of Creation," *Sojourners*, May 1977, pp. 23-26.

Chapter Six

[1]Alvin Toffler, *Future Shock* (New York: Bantam, 1970), p. 198.

[2]Paul Ramsey, *Fabricated Man* (New Haven, Conn.: Yale University Press, 1970).

[3]David Rorvick, *In His Image: The Cloning of a Man* (Philadelphia: Lippincott, 1978).

[4]"A False Image," *Newsweek*, 12 June 1979, p. 83.

[5]Peter Gwynne, "All About Clones," *Newsweek*, 20 March 1978, pp. 68-69.

[6]J. D. Broomhall, "Cloning Research," *Newsweek*, 17 April 1978, pp. 15-16.

[7]"Rorvick: Still Cloning Away," *Newsweek*, 14 January 1980, p. 17.

[8]"Clone Deemed a Hoax," *Dallas Times Herald*, 22 March 1981, p. 2-F.

[9]Matt Clark, "Clones Again," *Newsweek*, 12 February 1979, p. 99.

[10]Robert Briggs and Thomas J. King, "Transplantation of Living Nuclei

from Blastula Cells into Enucleated Frog Eggs," *Proceedings of the National Academy of Sciences* 38 (May 1952): 455–63.

[11]J. B. Gurdon, "Adult Frogs Derived from the Nuclei of Single Somatic Cells," *Developmental Biology* 4 (1962): 256–70.

[12]Lawrence E. Karp, *Genetic Engineering: Threat or Promise?* (Chicago, Ill.: Nelson-Hall, 1976), pp. 201–3.

[13]Robert Gilmore McKinnel, *Cloning: A Biologist Reports* (Minneapolis, Minn.: University of Minnesota Press, 1979), pp. 50–77.

[14]Harold M. Schmeck, "Study of Cloning Is Called Aid to Medical Research," *The New York Times*, 1 June 1978, p. A16.

[15]Joshua Lederberg, "Experimental Genetics and Human Evolution," *The American Naturalist* 100 (September–October 1966): 519–31.

[16]Bernard David, "Prospect for Genetic Intervention in Man," *Science* 170 (18 December 1970): 1282.

[17]Anthony Smith, *The Human Pedigree* (Philadelphia: Lippincott, 1975), pp. 52–54.

[18]J. B. S. Haldane, "Biological Possibilities in the Next Ten Thousand Years," in *Man and His Future*, ed. Gordon Wolstenholme (Boston: Little, Brown, 1963), p. 352.

[19]Joshua Lederberg, "Unpredictable Variety Still Rules Human Reproduction," *The Washington Post*, 30 September 1967.

[20]D. S. Halacy, Jr., *Genetic Revolution* (New York: Harper and Row, 1974), p. 163.

[21]McKinnell, *Cloning*, p. 104.

[22]Matt Clark, "Double Take," *Newsweek*, 12 March 1979, p. 97.

[23]Lederberg, "Experimental Genetics."

[24]Gurdon, "Adult Frogs."

[25]I. Scott Bass, "Governmental Control of Research in Positive Eugenics," *University of Michigan Journal of Law*, Spring 1974, p. 621.

[26]"Cloning," *The Macneil/Lehrer Report*, Show #3200, 7 April 1978, transcript pp. 3–4.

[27]*Los Angeles Times*, 17 May 1971, part iv, p. 1.

[28]N. N. Glatzer, *Hammer on the Rock: A Short Midrash Reader* (N.Y.: Schocken, 1962), p. 15.

Chapter Seven

[1]Herman Muller, "The Guidance of Human Evolution," in *Studies in Genetics*, ed. Muller (1962), p. 590.

[2]Anthony Smith, *The Human Pedigree* (Philadelphia: Lippincott, 1975), pp. 229–30.

[3]Jay Merwin, "How Far Should We Go With Human Engineering?" *Evangelical Newsletter*, March 1981.

[4]Joseph Fletcher, *The Ethics of Genetic Control* (Garden City, N.Y.: Doubleday, 1974), p. 199.

[5]Linus Pauling, "Reflections on the New Biology," *UCLA Law Review* 15 (1968): 269.

[6]Lori B. Andrews, "Embryo Technology," *Parents*, May 1981, p. 70.

[7]Charles Frankel, "Specter of Eugenics," *Commentary*, March 1974, p. 25.

[8]Richard Hutton, *Bio-Revolution: DNA and the Ethics of Manmade Life* (New York: New American Library, 1978), pp. 216–17.

Glossary

AMNIOCENTESIS—a medical procedure in which a needle is inserted into the amniotic cavity in order to withdraw amniotic fluid to diagnose particular genetic traits of the fetus.

CHROMOSOME—a linear arrangement of genes made up of DNA that is coated with protein. Chromosomes occur in pairs (called homologues) in all cells of an organism except an egg or sperm (humans have twenty-three pairs, or a total of forty-six).

DNA (deoxyribonucleic acid)—the double helix of nucleic acids that forms the molecular basis for heredity.

EUKARYOTES—organisms with a complex cell type containing a membrane-delimited nucleus that contains chromosomes.

GENE—the basic unit of inheritance; a particular sequence of DNA that codes for production of specific proteins.

GENOTYPE—the genetic constitution of an organism.

PHENOTYPE—the observable expression of the interaction of the organism's genetic structure and the environment.

PROKARYOTES—organisms with less complex organization (e.g., bacteria) composed of a simple circular DNA molecule that is not membrane-delimited.

RNA (ribonucleic acid)—any of the various nucleic acids that transfer the genetic code from DNA to protein manufacture.

Bibliography

Anderson, Bruce L. *Let Us Make Man*. Plainfield, N.J.: Logos International, 1980.

Anderson, Norman. *Issues of Life and Death*. Downers Grove, Ill.: InterVarsity, 1977.

Davis, Bernard et al. "Recombinant DNA Research: A Debate on the Benefits and Risks." *Chemistry and Engineering News*, 30 May 1977.

Ellison, Charles, ed. *Modifying Man: Implications and Ethics*. Washington, D.C.: University Press of America, 1977.

Hatfield, Charles, ed. *The Scientist and Ethical Decisions*. Downers Grove, Ill.: InterVarsity, 1973.

Howard, Ted, and Rifkin, Jeremy. *Who Should Play God?* New York: Dell, 1977.

Karp, Laurence E. *Genetic Engineering: Threat or Promise*. Chicago, Ill.: Nelson-Hall, 1976.

Kass, Leon. "The New Biology: What Price Relieving Man's Estate?" *Science* 174 (19 November 1970): 779–88.

Lester, Lane P., with Hefley, James C. *Cloning: Miracle or Menace?* Wheaton, Ill.: Tyndale, 1980.

Packard, Vance. *The People Shapers*. Boston: Bantam, 1977.

Ramsey, Paul. *Fabricated Man*. New Haven, Conn.: Yale University Press, 1970.

_____. "Shall We Reproduce?" *Journal of the American Medical Association* 220 (5 and 12 June 1972): 1346–50, 1480–85.